MINI/MICRO SOLDERING AND WIRE WRAPPING

MURRAY P. ROSENTHAL

RCA American Communications, Inc.

HAYDEN BOOK COMPANY, INC.
Rochelle Park, New Jersey

Library of Congress Cataloging in Publication Data

Rosenthal, Murray P
    Mini/micro soldering and wire wrapping.

    Includes index.
    1. Miniature electronic equipment -- Amateurs' manuals.
2. Solder and soldering -- Amateurs' manuals. I. Title.
TK9965.R66   621.381'7   78-6771
ISBN 0-8104-0864-3

Printed in the United States of America

1	2	3	4	5	6	7	8	9	PRINTING

78 79 80 81 82 83 84 85   YEAR

Preface

Walk into any electronics parts distributor or computer store and you can find any number of books on topics ranging from *Integrated Circuits* through *Servicing Your Own TV Set* to *How to Select and Use Hi-Fi and Stereo Equipment.* You can also pick up books on various projects you can build, from a simple code-practice oscillator to a microprocessor. But I defy you to find a book on one of the most important facets of all the foregoing subjects: how the individual electronic components are fastened together in the first place and/or how best to remove them during repair and troubleshooting.

This book attempts to fill that void. Its four chapters cover information on the three methods of fastening: fastening new circuits, unfastening/refastening existing circuits, and some basic techniques for troubleshooting and repairing electronic equipment.

Acknowledgments

I wish to thank the following for their help and cooperation in the preparation of this book: Antex Div., M.M. Newman Corp.; Belmont Metals, Inc.; Caig Laboratories, Inc.; EREM Corp.; Hewlett-Packard; Hexacon Electric Co.; Hollis Engineering, Inc.; Hunter Tools; JERMYN Corp.; Kester Div., Litton Systems, Inc.; Klein Tools, Inc.; Lintek Co.; MOSTEK Corp.; OK Machine & Tool Corp.; Pomona Electronics; RCA Corp.; Sonobond Corp.; Teledyne Kinetics; Triton Mfg. Co.; Ungar Div., Eldon Industries; Weller Div., Cooper Group; Wen Products, Inc.; Zeva Electric Corp.; Doreen Keyes, who somehow typed sense out of my scribblings; and my wife Rita, who managed to stifle her screams when she looked at the mess I made of our playroom/office.

Contents

1
Introduction

When one thinks of electronic equipment, one pictures a maze of wiring connecting together all sorts of devices ranging from diodes less than ¼ inch long through tube sockets to somewhat massive transformers. Although this is a true picture of the older pieces of electronic gear that still exist, the picture of today's gear is somewhat cleaner. Whether it be computer, microprocessor, hi-fi rig, or TV set, the only wiring one usually sees is that running between one printed-circuit (PC) board and another, between boards and controls, between a picture tube and a board or boards, or between plug-in connectors that accommodate PC boards. Any other wiring is on the boards themselves and consists of a pattern of conductive material (copper, almost exclusively) applied to one or both sides of an insulating substrate.

Many reasons exist for the use of printed rather than hard-wired circuits:

1. Most important is the saving in space that is engendered.
2. A complex circuit may be modularized by using several small printed circuits instead of a single larger circuit. Such modularity simplifies troubleshooting, circuit modifications, and mechanical assembly within any particular housing.
3. Soldering of component leads may be accomplished in an orderly sequence by hand (the basis for all of today's kits) or simultaneously during a production cycle.

4. A more uniform product results because wiring errors are eliminated and because distributed capacitances are constant from one production unit to another.
5. The printed-circuit method of construction is more amenable to automatic assembly and testing equipment.
6. Printed circuits can be "layerized" in a sort of sandwich, with the respective conductors separated by insulating layers.
7. If the appropriate substrate is used, the printed circuits can be made flexible.
8. While most printed circuits are thought of as mere boards containing the printed wiring connecting the components on the boards themselves, the components themselves can also be printed. Integrated circuits (ICs) are a prime example of this technique (Fig. 1-1), but capacitors can also be produced by

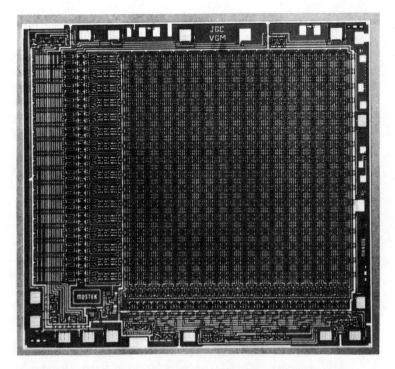

Fig. 1-1. Large-scale integration (LSI), 1024-bit dynamic random-access memory. *(Courtesy* MOSTEK Corp.)

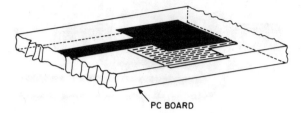

Fig. 1-2.   Printed-circuit capacitor.

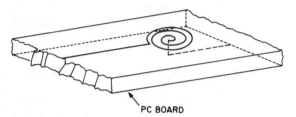

Fig. 1-3.   Printed-circuit spiral inductor.

printing conducting areas on opposite sides of the PC board using the board itself as the dielectric (Fig. 1-2). Furthermore, spiral-type inductors can be printed (Fig. 1-3).

Definitions

Today's electronic circuits, whether printed or hard-wire, are connected in either of two fashions: soldering or wire wrapping. It will be helpful if both terms are properly defined.

Wire Wrapping

Wire wrapping is a specialized solderless process of tightly coiling a solid conductor wire around a rigid metal terminal, as shown in Fig. 1-4, for the purpose of electrical connection. Wire wrapping is not new. The principle was developed in the early 1950s by the Bell Laboratories for Western Electric Company, which had found production soldering of telephone equipment to be an increasingly unworkable connection technique in the face of growing connection volume, contact density, and product thermal sensitivity. The technique was seriously adapted in the mid-1950s, and today it accounts for approximately 98 percent of the discrete interconnects in the

Fig. 1-4. Wire-wrapped terminal. *(Courtesy* OK Machine & Tool Corp.)

United States telephone industry. During the 1960s the technique was adopted by the electronic field, and today it is used throughout the electronics, telecommunications, and electrical products industries throughout the world.

Basically wire wrapping is a production technique, employing both automatic machinery (Fig. 1-5); and/or hand-wrapping tools, such as those shown in Figs. 1-6 and 1-7. However, it does come into the world of the individual during research and development, where prototype units are put together; in troubleshooting and repair of existing wire-wrapped equipment; and in the rapidly burgeoning field of home microcomputer kits.

Soldering

Basically, soldering (per the American Welding Society) is a joining of metals by adhesion using a non-ferrous metallic bonding

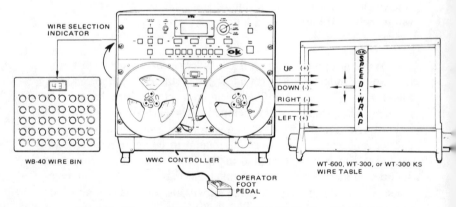

Fig. 1-5. Wire-wrapping machinery. *(Courtesy* OK Machine & Tool Corp.)

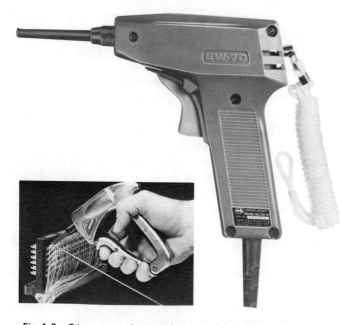

Fig. 1-6. Trigger-type wire-wrapping tools. (A) Model EW-7D electric-powered tool; (B) Model G-200/R3278 manual tool. *(Courtesy* OK Machine & Tool Corp.)

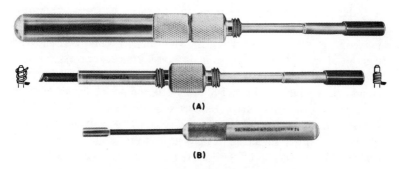

Fig. 1-7. Hand-wrapping tools. (A) Model HW-UW224 wire-wrapping/wire-unwrapping tool; (B) Model HW-26 wire-wrapping tool. *(Courtesy* OK Machine & Tool Corp.)

alloy. The bonding or soldering alloy has a relatively low melting point, below 800°F (427°C), and is preferably present as a *thin* film between the parts to be joined. To obtain good adherence and strength, most metals require surface preparation—usually through the application of a flux to remove oxides from the surface. In electronic soldering, the flux is usually the core of the solder itself, that is, the solder is essentially a soft, flexible wire whose core is filled with a material (e.g., resin), that melts before the metal does, flowing onto and preparing the surfaces to be joined.

The most important requirement for a soldered joint is that of metallic continuity between the solder and the surfaces being soldered. The atoms of the solder actually form metallic bonds with the atoms of the metals being soldered. The alloy layer may be only a few atoms thick in some cases, while in other instances an appreciable thickness of an alloy may be present. In order to attain this unity between the solder and the metals being joined, both the joint and the solder must be heated to a temperature sufficient to melt the solder and raise it to soldering temperature.

In addition to heat, continuity depends on the "solderability" of the metals being joined. By solderability is meant the ease with which molten solder will "wet" the surfaces of the metals being joined. Wetting is a surface phenomenon which depends, among other factors, primarily on surface cleanliness. When molten solder leaves a continuous *permanent* film on a metal surface, it is said to have "wet" the surface. Wetting is facilitated by fluxing, and solderability depends on the activity of flux.

Heat for melting the solder and its flux and for warming the parts to be joined is usually applied by means of the familiar soldering iron, variations of which are shown in Figs. 1-8 and 1-9. The iron has the advantage of always being on, so that the heat necessary is always available. This also has its disadvantages: Unless a protective stand is available (see Figs. 1-8 and 1-10), the possibility of burning the user and/or work surface is ever present, to say nothing of the constant drain on one's source of electricity. To offset these disadvantages, other soldering devices have been developed. All these tools shown in Figs. 1-11, 1-12, and 1-13 remain off until triggered into action; the heat then comes up in a matter of a very few seconds. Such devices need no protective stands and are on only as long as they are in actual use.

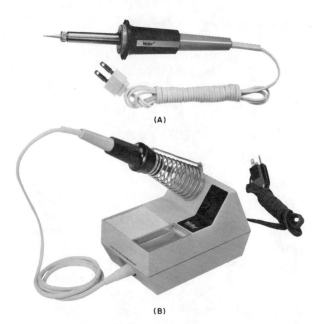

Fig. 1-8. Fixed-element soldering irons. (A) Simple fixed-element iron; (B) fixed-element iron in powered stand; the stand also contains a sponge type of wiper/cleaner for the tip and a receptacle for miscellaneous items. *(Courtesy* Weller Div., Cooper Group)

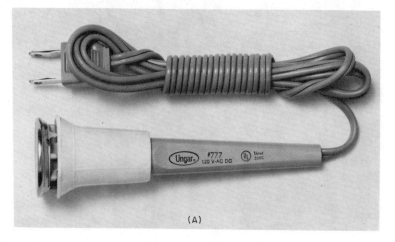

Fig. 1-9. Model 777 variable-element/tip soldering iron. (A) Power and handle unit. (*Courtesy* Ungar Div. Eldon Industries)

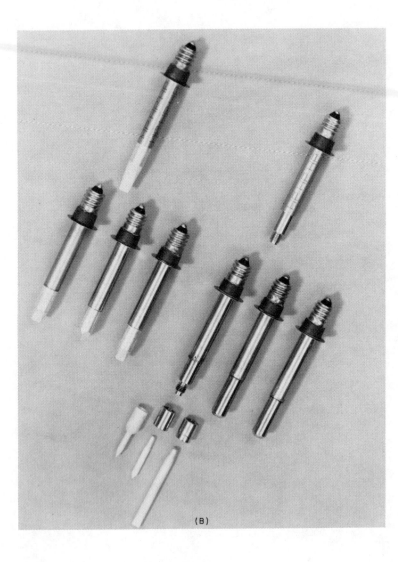

(B)

Fig. 1-9   Model 777 variable-element/tip soldering iron. (B) Element and
tip units. (*Courtesy* Ungar Div. Eldon Industries)

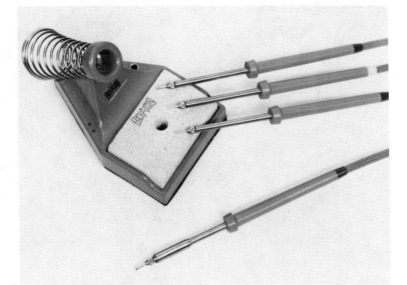

Fig. 1-10. Model 8965 micro-mini holder, sponge, and irons. (*Courtesy* Hexacon Electric Co.)

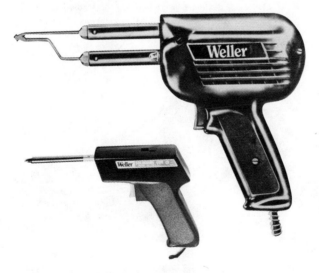

Fig. 1-11. Weller soldering guns. (A) Standard gun with automatic spotlight; (B) Tempmatic ® Model GT-7A/6B single-post gun. (*Courtesy* Weller Div., Cooper Group)

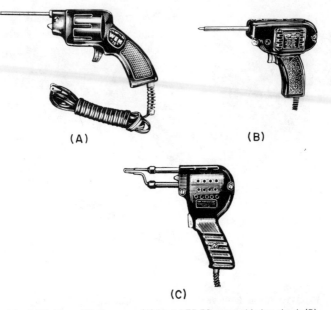

(A)   (B)

(C)

Fig. 1-12. Wen soldering guns. (A) Model 75 50-watt soldering pistol; (B) Model 222 dual-heat, single-post gun; (C) Model 250 250-watt standard gun. All have automatic spotlights. (*Courtesy* Wen Products, Inc.)

Fig. 1-13. Model J Pres-to-Heat soldering tool. Depressing the trigger in the handle slightly clamps tool jaws around work; depressing the trigger further will activate the on-off switch, causing tool jaws to heat. (*Courtesy* Triton Mfg. Co.)

Production soldering is another story. If manufacturers still had to rely on individuals soldering the components onto discrete PC boards, consumer costs would be sky-high. Thus, mass soldering has come into being, the most prevalent being wave soldering, with other techniques also available, such as dip, drag, and ultrasonic fluxless soldering.

Wave Soldering

When molten solder is pumped upward from a reservoir through a nozzle, a standing wave of solder is formed. Electronic components on PC boards passed over the wave (Fig. 1-14) are soldered to the circuit. Many such connections can be made in one pass of the PC board over the solder wave, whereas in hand soldering only one connection at a time is soldered. A typical wave-soldering system consists of applying flux to the printed circuit, preheating the circuit board, and performing the wave-soldering operation. The flux

Fig. 1-14. Adjustable-width finger conveyer carriers PC board assembly over deep, wide, stable solder wave. *(Courtesy* Hollis Engineering, Inc.)

mechanism applies flux uniformly to the side of the PC board on which the soldered connections will be made. Plated-through holes require flux throughout the hole to prepare for soldering through-hole connections to circuits on the other surface of the PC board when it hits the molten solder wave. The preheating equipment has several functions. Preheating reduces the thermal shock to the board when it hits the wave of molten solder. Preheating also drives off the bulk of the flux solvent, so that the board accepts the solder more readily. Finally, preheating permits the flux to rise more readily into the plated-through holes, thus performing its deoxidizing action more completely. The solder module itself not only does the soldering, it removes excess solder so that no solder "bridges," "icicles," or other globs of solder are formed.

In a sophisticated wave-soldering system, such as that shown in Fig. 1-15, a process termed *stabilizing* is also included. When the components are first inserted into the PC board, the leads extend down, sometimes as much as 2 inches, in all sorts of directions. These leads must be cut off as close to the board as possible, still leaving them long enough to take solder. However, with the leads hanging loosely, lead trimming cannot be uniform. Therefore, the board must be passed through some sort of nonconducting fixative, usually melted refined petroleum wax. The melted wax adheres to

Fig. 1-15. Stabilizer 160 system carries boards from assembly line through the stabilizing process, lead cutting, and wave soldering. Conveyorized cleaning unit using water completes the operation. *(Courtesy* Hollis Engineering, Inc.)

the leads and when cooled solidifies rapidly, locking the leads in place. The locked leads are now stabilized, that is, steady enough to go through high-speed carbide cutters and come out just long enough for soldering.

Dip Soldering

Dip soldering is exactly what the name implies: The work (PC board) to be soldered is dipped into a container of molten solder, thereby effecting the necessary connections. Herein lies the basic difference between dip and wave soldering: Dip soldering calls for immersing the board into the solder, whereas wave soldering calls for the solder to come up to the board. Otherwise, the systems are similar, both using fluxers, preheaters, and soldering machines. One dip-soldering arrangement is pictured in Fig. 1-16. This is a manual operation, with human operators (or one operator alone) performing all three functions.

Drag Soldering

Drag soldering is similar to wave soldering in that both use conveyors to move the PC boards to be soldered. The similarity ends there, however. In drag soldering, the PC board moves (is dragged) through the molten solder in one direction, while the solder flows in the opposite direction.

One such drag-soldering system is the Zeva TSM System (Fig. 1-17). The PC board is placed into the carrier and fed into the machine, where it is automatically picked up by the conveyor mechanism. The board then travels over the fluxer, which sprays a foamed flux on the underside of the board. The board then passes over a skimmer blade, which trims off any excess solder, and travels thence to the predryer, where the flux is dried by means of infrared heating. Following this, the machine automatically tilts the board holder to a slight angle and drags the board through the molten solder. The angle permits the excess solder to flow back into a receiving tank, where any flux or dross in the solder is removed before the solder is returned to the molten solder bath. The completed board is then conveyed onto the exit rails, ready for removal of component leads. Depending on the Zeva machine used, component lead lengths of up to 1 inch can be accommodated. The entire process can be automatic or controlled manually.

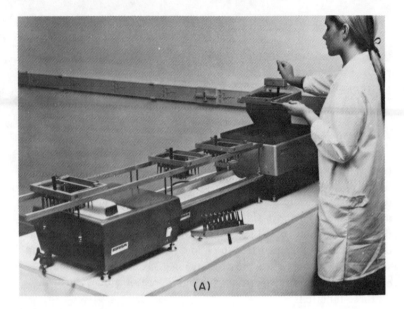

Fig. 1-16. ERSA dip-soldering system. (A) In-line production with F3380 fluxer, H3370 preheater, and T3300 soldering machine. (B) T3300 soldering machine. *(Courtesy* Caig Laboratories, Inc.)

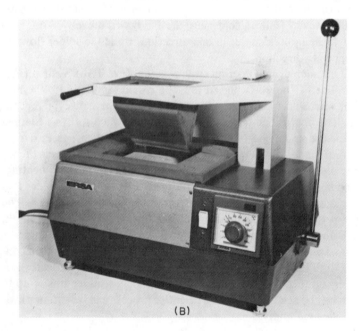

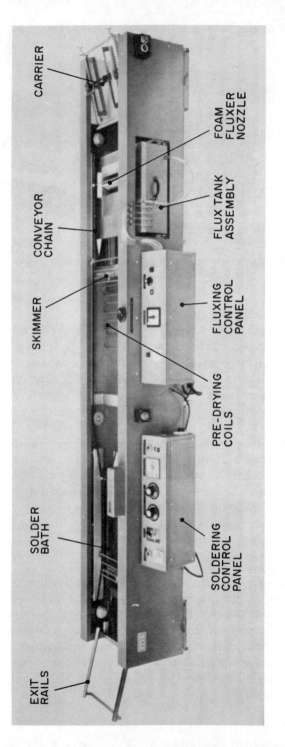

CARRIER

FOAM FLUXER NOZZLE

FLUX TANK ASSEMBLY

CONVEYOR CHAIN

SKIMMER

FLUXING CONTROL PANEL

PRE-DRYING COILS

SOLDER BATH

SOLDERING CONTROL PANEL

EXIT RAILS

Fig. 1-17. Zeva TSM/FT drag-soldering system. *(Courtesy Zeva Electric Corp.)*

15

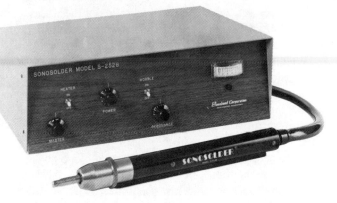

Fig. 1-18. Sonosolder Model S-2528 ultrasonic soldering system. *(Courtesy* Sonobond Corp.)

Ultrasonic Soldering

Ultrasonic soldering is a method whereby nonferrous metals can be soldered without the need for a flux. The method can be used in either single-joint soldering or production-line, multijoint soldering of such metals as copper, brass, aluminum, silver, tantalum, silicon, and germanium.

The principle is simple: Ultrasonic vibrations are used literally to "tear" any oxides away from the surface(s) to be soldered (or tinned), obviating the need for flux. The unit illustrated in Fig. 1-18 is both a 25-watt soldering iron and an ultrasonic vibrator whose frequency of vibration is 28 kHz (kilohertz). The iron itself will accept an ultrasonic probe, which can be used to vibrate a bath of molten solder and so provide flux-free dip soldering. Or it will accept various sizes of soldering tips, from 0.005 inch on up, for standard soldering/tinning tasks.

Brazing and Welding

Brazing, as defined by the American Welding Society, is a group of welding processes wherein the filler metal is a nonferrous metal or alloy whose melting point is higher than 1,000°F (538°C)but lower than that of the metals or alloys to be joined. At one extreme, brazing is very similar to soldering and is sometimes called *hard soldering*. A typical example of this is silver soldering, which uses alloys containing 10-80% silver, with the balance being mainly copper and zinc. The melting points of these alloys are in the range of 1,175–1,500°F (635–815°C), which is considerably higher than that for soft solders.

Welding is a method of joining metals by means of fusion or by interdiffusion in the solid state under the influence of high temperature and pressure. Metals with similar composition may be united into one homogeneous piece by fusing together the edges in contact, or by the addition of molten metal of the proper characteristics deposited where it will form a fused joint with each piece.

Since neither brazing nor welding is of any real concern in electronic circuits, the topics can be dropped right here. The definitions were included only to point out the differences between soldering, brazing, and welding.

Solders

Soldering alloys include a wide range of alloying elements and properties. The common characteristic of these alloys is that they are used almost entirely for joining purposes and that they have a liquidus (lowest temperature at which they are completely liquid) of less than 800°F (427°C), Table 1. Practically all solders used for electrical connections melt below 625°F (329°C).

Solder Alloys

Tin-Lead: This is the largest single group and the most widely used of the soldering alloys. Most metals can be joined with these alloys. Tin-lead solders are compatible for use with all types of base metal cleaners, fluxes, and heating methods.

The most widely used solders—often referred to as general-purpose solders—of the tin-lead system are the classifications 35A through 50A. They provide optimum wetting and flow properties, good strength at low temperatures, and require less preparation time.

The 60A solder, almost the eutectic (low melting), is particularly adaptable to delicate work or when soldering temperature may be critical. The solder with the highest tin content, 70A, is used for soldering zinc.

Tin-Lead-Antimony: These alloys are generally used in the same applications as the tin-lead solders. They are not recommended for use on aluminum, zinc, or galvanized steel, however. The addi-

Table 1. Typical Solders.

ASTM Class.		Composition (percentage by weight)			Temperature (F) Solidus	Liquidus	Pasty Range
		Tin	Lead				
5A		5	95		572	596	24
10A		10	90		514	573	59
15A		15	85		437	553	116
20A		20	80		361	535	174
25A		25	75		361	511	150
30A		30	70		361	491	130
35A		35	65		361	477	116
40A		40	60		361	455	94
45A		45	55		361	441	80
50A		50	50		361	421	60
60A		60	40		361	374	13
70A		70	30		361	378	17
	Tin	Antimony	Lead				
20C	20	1.0	79.0		363	517	154
25C	25	1.3	73.7		364	504	140
30C	30	1.6	68.4		364	482	118
35C	35	1.8	63.2		365	470	105
40C	40	2.0	58.0		365	448	83
	Lead	Silver	Tin				
2.5S	97.5	2.5	. .		579	579	0
5.5S	94.5	5.5	. .		579	689	110
1.5S	97.5	1.5	1.0		588	588	0
		Tin	Silver				
		96.5	3.5		430	430	0
		95	5		430	473	43
		Tin	Zinc				
		91	9		390	390	0
		80	20		390	518	128
		70	30		390	592	202
		60	40		390	645	255
		30	70		390	708	313
		Cadmium	Zinc				
		82.5	17.5		509	509	0
		40	60		509	635	126
		10	90		509	751	241
Tin	Indium	Bismuth	Lead	Cadmium			
8.3	19.1	44.7	22.6	5.3	117	117	0
12	21	49	18	. .	136	136	0
12.8	4	48	25.6	9.6	142	149	7
50	50	. .	. .	. .	243	260	17
48	52	. .	. .	. .	243	243	0

tion of antimony up to a maximum of 6% of the tin content does not seriously affect the wettability or flow characteristics, but it does increase the mechanical properties.

These solders have almost the same properties as equivalent tin-lead solders with about 5% more tin. The physical properties—tensile strength, creep (slow, plastic flow under pressure), and hardness—are higher than the nonantimonial alloys.

Tin-Antimony: The 95% tin, 5% antimony solder has the best electrical properties of any solder alloy. In addition, it has high strength at temperatures to 300°F (149°C), and it has excellent flow characteristics.

Tin-Silver: The tin-silver alloys exhibit the same characteristics as the tin-antimony solders and are used for delicate instrument work.

Tin-Zinc: This group of alloys is not considered highly important for soldering applications. Its widest use is for soldering aluminum—primarily where lower soldering temperature than that for zinc-aluminum solders is required.

Lead-Silver: The physical properties of these solders—tensile strength, creep, and shear strengths—are good to 350°F (177°C). Their fatigue properties are also better than those of the nonsilver solder alloys. The silver-lead solders, however, have poor wetting characteristics and are subject to corrosion when stored in a humid atmosphere. The addition of 1% tin, replacing silver, increases wetting and flow and reduces corrosion susceptibility.

Cadmium-Zinc: The cadmium-zinc solders are useful in joining aluminum to itself or other metals. It is particularly useful in applications where service temperature is higher than that allowed for use with alloys of lower melting points.

Zinc-Aluminum: Designed specifically for soldering aluminum, zinc-aluminum solders provide high joint strength with good corrosion resistance. A 95% zinc, 5% aluminum solder has been used without flux in electronics applications.

Indium Solders: Indium solders are used only in special applications—for instance, in cryogenic applications. The 50% indium, 50% tin alloy is particularly useful in glass-to-glass and glass-to-metal soldering.

Fusible Alloys: These bismuth-containing solders are used when soldering must be performed at temperatures below 361°F

(183°C). They are more sensitive to long periods of loading above room temperature (creep). The higher bismuth-content solders are not easily adaptable to high-speed soldering operations, nor do they easily wet the base metal.

Solder Forms and Shapes

Soldering alloys are commercially available in practically any desired shape, weight, or size. Typical examples of available forms are: pig, slab, cake, or ingot (rectangular or circular in shape); bar, paste, ribbon, or tape (thicknesses of $\frac{1}{16}$-$\frac{3}{16}$ inch); segment or drop (triangular bar or wire cut into pieces); pulverized or powdered; foil (thicknesses from 0.00125 inch); sheet (thicknesses of 0.010-0.100 inch); solid wire; flux core wire; and preforms (Fig. 1-19) such as rings, washers and disks, pellets, spheres, sleeves, wire forms, and special stampings.

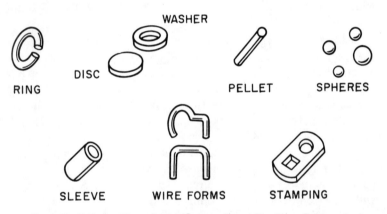

Fig. 1-19. Typical solder preforms. (*Courtesy* Kester Div., Litton Systems, Inc.)

2

Wiring and Soldering New Circuits

Assuming you are not in the business of producing new wired/printed circuits on a production basis, there are basically only two types of new circuits that require soldering: permanent and breadboard. Breadboard circuits are almost always fashioned for temporary purposes, such as putting together and testing a new design. Because this book emphasizes putting together permanent circuits, we will confine the discussion to them. Besides, what is true for permanent circuits is true in general for breadboards, except that you need not make your breadboard connections as permanent—you have to undo them later.

Hard-Wired Circuits

Have you ever seen the wiring of military or high-class industrial electronics equipment and wondered at the precision construction? Such equipment can be packed with components, wire, and hardware, and yet seem to have a place for everything. Service points are usually easily accessible even in the most compact chassis. There can be a hundred wires or more, but to the untrained eye there appear to be only a few cables and harnesses here and there. This type of electronic construction has been well thought out and results from a team effort of engineers, technicians, and assembly-line workers.

It would be nearly impossible for you to duplicate this type of construction when building an individual piece of equipment (unless, of course, you're working from a set of kit instructions). However, there are certain techniques and suggestions that can be followed to produce superior results.

Tools Required

Whatever the type of wiring to be done, there are several tools required. Table 2 lists this complement in descending order of im-

Table 2. Required Wiring Tools.

	Tool	Use
1.	Screwdriver set (5) (1/8-in. through 5/16-in.)	Mounting/removing wire-connection hardware
2.	Subminiature long-nose pliers	Wiring/unwiring in tight quarters and/or general wiring
3.	Duckbill pliers	Same as Item 2; also for holding nuts while mounting hardware
4.	Phillips-head screwdriver set (1/8-in. through 5/16-in.)	Same as item 1 for Phillips-head screws
5.	Subminiature cutters	For cutting wiring in tight quarters and/or cutting small-gauge wire to required lengths; see text for other uses.
6.	Standard electronic cutters	For cutting and/or stripping heavier gauge wiring
7.	Gas pliers	For various heavy-duty tasks, such as tightening the nuts on controls and switches being mounted on the chassis

The following are not mandatory but are most helpful:

8.	Wire stripper	For quick removal of wiring insulation
9.	Socket wrench set or nut-driver set (5/32-in. through 5/16-in.)	For holding/tightening nuts; see text for other uses
10.	Open-end wrenches (7/16-in. and 1/2-in.)	For tightening nuts on controls and switches
11.	Wire-wrap tool	For use when wire wrapping is desired vs. soldering
12.	Crimping tool	For crimping terminal lugs on wire ends when connections to screws or spade-type connections are required

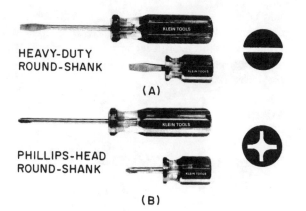

HEAVY-DUTY
ROUND-SHANK

(A)

PHILLIPS-HEAD
ROUND-SHANK

(B)

Fig. 2-1. Required screwdrivers. (A) Standard screwdriver and screw head; (B) Phillips screwdriver and screw head. *(Courtesy* Klein Tools, Inc.)

Fig. 2-2. Required pliers. (A) Long-nose; (B) duckbill; (C) cutting; (D) gas. *(Courtesy* Klein Tools, Inc.)

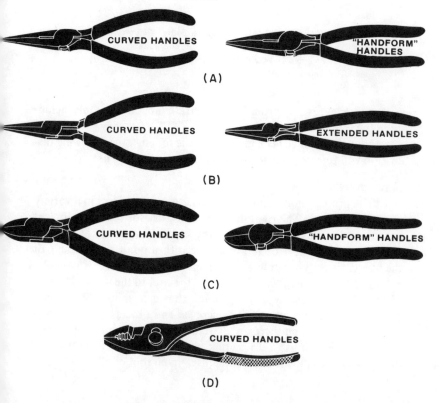

CURVED HANDLES   "HANDFORM" HANDLES

(A)

CURVED HANDLES   EXTENDED HANDLES

(B)

CURVED HANDLES   "HANDFORM" HANDLES

(C)

CURVED HANDLES

(D)

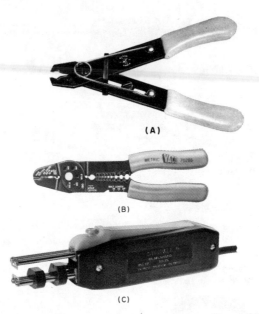

Fig. 2-3. Wire strippers. (A) Wire stripper/cutter. *(Courtesy* Klein Tools, Inc.) (B) Wire stripper/cutter and bolt cutter. *(Courtesy* Vaco Products Co.) (C) Stripall thermal wire stripper. *(Courtesy* Teledyne Kinetics)

portance. Figures 2-1 through 2-5 illustrate most of the tools included in the table.

Now that you've been given the basic tool list, the following discussion will refresh your memory as to their care and use.

Screwdrivers

The professional requires a number of screwdrivers in a variety of sizes and types. The right driver is necessary for the fast, efficient driving and removal of screws in any kind of material. The wrong size driver—too short or too long—or with a point that doesn't fit the screw properly—can waste time and make trouble.

The basic rule, of course, is to fit the tool to the work. The size of the screw and the type of opening it has determine which driver you use. Slotted screws, Phillips screws, and others come in many sizes and lengths, and there are drivers to fit all of them. There are a few tips on *how* to use a driver that we can all benefit from because the screwdriver is the most often abused hand tool of all.

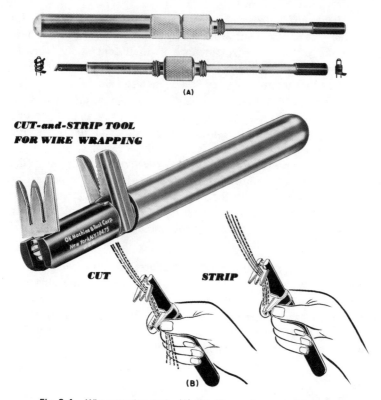

Fig. 2-4. Wire-wrapping tools. (A) Hand-type wire wrapper/unwrapper with removable cap; (B) Hand-type wire cutter/stripper. *(Courtesy* OK Machine & Tool Corp.)

Fig. 2-5. Crimping and stripping tool. *(Courtesy* Klein Tools, Inc.)

- *Never use a driver to do another tool's job.* Using a driver as a pry bar or scraper or chisel can ruin the tool and spoil the work, not to mention the time you might lose or the injury you might sustain. Limit screwdrivers to screws.

- *Never push a driver beyond its capacity.* Make a pilot hole for the screw, and your work will go more easily. If you can't make a hole, at least prepunch a small dent in the work to guide the screw point. Use a square-shank driver for heavy work that requires the use of a wrench to help do the turning.
- *Never expose a driver to excessive heat.* Direct flame can draw the temper (hardness) from the metal, weakening and possibly warping it, making it unsafe and inefficient to use. If the handle is made of plastic, it may melt.
- *Never use a driver at an angle to the screw.* Always keep the shank perpendicular to the screw head. This is as important as using a point that fills the screw opening. Driving at an angle to the screw or using a point that is too small can spoil the screw and also make the point slip and damage the work.

Attempts to repair most types of drivers are *not recommended*. The tips of drivers for slotted screws can be dressed on a bench grinder, but care must be taken not to let the tip get too hot. (If it gets hot to the touch, the temper has been drawn, and the tip will no longer stand up well.) Drivers with cracked handles, bent or twisted shafts, or worn tips should be discarded and replaced.

Pliers

Each type of pliers does its own particular job better than any other type. Choosing the right pliers helps you do the job at hand more easily, more efficiently, and more safely.

The professional takes his job and his tools seriously. He will equip himself with every kind of pliers he needs because he knows this small investment will pay off continuously in saved time and professional results.

The professional knows all the rules about using pliers. He has heard them or read them or learned them through experience. If the following list helps you to recall even one rule you may have forgotten, then it will be more than worthwhile.

- *Never use pliers to do another tool's job.* A pair of pliers is not a hammer or a pry tool or a wrench. Using pliers instead of the proper tool risks damaging the pliers, damaging the work, damaging yourself, and losing time. It's just not worth it.

- *Never push pliers beyond their capacity.* Bending stiff wire with light pliers or the tip of needlenose pliers can spring them or break them. Use a stronger pair with a blunt nose. When you need greater leverage, use pliers with greater leverage. Don't extend the length of the handles. Bolts should be cut with a bolt cutter; large cables should be cut with a cable cutter. To each its own.
- *Never expose pliers to excessive heat.* Direct flame on metal can draw the temper and ruin the tool. Cutting pliers are especially vulnerable to high, direct heat.
- *Never cut hardened wire with ordinary pliers.* Pliers should not be used for cutting hardened wire unless they are recommended specifically for this use.
- *Never rock pliers from side to side when cutting wire and never bend the wire back and forth against the cutting edges.* Either practice can dull or nick the cutting edges. Cut wire at a right angle only. If the wire won't cut through readily, the cutting edges may need sharpening. Or you may need pliers with greater leverage.
- *Never cut any wire or metal unless your eyes are protected.* Safety glasses or other protective devices are a must. It's easy to forget to wear them. It's a big bother to put them on for "just one cut." You've heard all the excuses and reasons, but none of them make any sense. They're all part of the lazy man's way, not the professional's way—the safe way!
- *Never depend on plastic-dipped handles to insulate you from electricity.* Ordinary plastic-dipped handles are there for comfort and a firmer grip. They are *not* intended for protection against electric shock.

Attempts to repair pliers are *not recommended.* However, proper maintenance is a must. Dull cutting edges can be sharpened with a small, medium-grade honing stone. The knurl inside the nose of the pliers may be cleaned with a wire brush. An occasional drop of oil at the hinge will help to preserve the pliers and assure easy operation. Pliers that may not be used for a long stretch should be cleaned with a degreaser (take care not to dissolve any plastic coatings), then wiped with an oil-soaked rag before storing. Any pair of pliers that is cracked, broken, sprung, or has nicked cutting edges should be discarded and replaced.

Wrenches

Whatever the job, only a wrench of the proper type and size will give you the kind of results you want. That one right wrench will do the job correctly, with less effort and more safety than any other wrench.

Take a minute to read these few simple rules; you might find something you've almost forgotten.

- *Never use a wrench to do another tool's job.* You won't do the job as well, and you might damage or even break the wrench. Using a wrench as a hammer or a pry bar or anything else can be dangerous. Take the time to get the right tool.
- *Never use a wrench opening too large for the fastener.* Using a wrench opening too large for the nut or bolt can spread the jaws of an open-end wrench and batter the points of a box or socket wrench. An opening that is too large can also spoil the points of the nut or bolt head. When selecting a wrench for proper fit, take special care to use inch wrenches on inch fasteners and metric wrenches on metric fasteners.
- *Never push a wrench beyond its capacity.* Quality wrenches are designed and sized to keep leverage and intended load (torque) in safe balance. The use of an artificial extension on the handle of any wrench can break the wrench, spoil the work, and hurt the user. Instead, get a larger wrench or a different type of wrench to do the job. The safest type is a box or socket wrench. To free a "frozen" nut or bolt, use a striking-face box wrench or a heavy duty box or socket wrench. Never use an open-end wrench. Always apply penetrating oil beforehand.
- *Never expose a wrench to excessive heat.* Direct flame can draw the temper from the metal, weakening and possibly warping it, making it unsafe to use.
- *Never push on a wrench unless absolutely necessary.* There may be situations in which you can only *push* a wrench handle to loosen or tighten a nut or bolt. But *you should always pull on a wrench* to exert even pressure and avoid injury if the wrench slips or the nut breaks loose unexpectedly. If you must push the wrench, do it with your palm and hold your hand open.
- *Never cock or tilt an open-end wrench.* Always be sure the nut or bolt head is fully seated in the jaw opening for both safety

and efficiency. A box or socket wrench should be used on hard-to-reach fasteners. Adjustable wrenches should be adjusted tightly to the work and pulled so that the work is applied to the *fixed jaw.*

- Never depend on plastic-dipped handles to insulate you from electricity. Ordinary plastic-dipped handles are there for comfort and a firmer grip. They are *not* intended for protection against electric shock. Special high-dielectric handle insulation is available, but even this should be used only as a secondary precaution.

Attempts to repair box, open-end, or combination wrenches are *not recommended.* Any of these wrenches with bent handles, spread, nicked, or battered jaws, or rounded or damaged box points should be discarded and replaced.

Socket and adjustable wrenches can be repaired by the replacement of the damaged parts. Periodic inspection, cleaning, and light lubrication will maintain these wrenches and reveal any damage. An adjustable wrench with a spread or damaged fixed jaw or a bent handle should be discarded and replaced. Bent socket wrench handles and extensions and cracked or battered sockets should be discarded and replaced.

Parts Placement

There are many factors to consider when you're deciding where to place the parts for your particular project. Among them are such items as chassis design, front panel layout, wiring simplification, test point accessibility, the number of internal parts, and the location of direct alternating current (ac) components (i.e., transformers, lamps, auxiliary sockets, etc.). The final parts placement is usually a compromise solution resulting from careful consideration of all these items.

To some degree the kind of chassis determines the placement of parts. Large items may fit in only one area of the chassis. The way the case is assembled and disassembled governs many of the positions of the parts. The easiest equipment to wire is that which is relatively open, with room for your hands to work, and which doesn't require cables between sections of the chassis. If you are building a console type of cabinet with a sloping front panel, fasten all the parts to the front panel or to a subchassis that is connected to the front panel. Then, when the back and side sections are removed, there will be no parts or wires connected to them.

The front panel should be designed for ease of operation, uniform component placement, and pleasing appearance. Meters are usually placed highest so they are easier to read. Try to place jacks close to the bottom so that their leads will not interfere with or cover up any other front panel component. If there are a number of switches, try to position the switches left to right in order of their use. If you use lamps to indicate switch positions, keep the lamps close to their respective switches, preferably directly above them. Fuses should also be made accessible via the front panel.

If you are making a multipurpose instrument, keep together all the controls for each section. Another factor that is often overlooked is knob size. Be sure to leave enough room between front panel components for the knobs, calibration marks, and finger space. It's very irritating to try to turn a knob or flip a switch and not have enough space for your fingers.

Mount the internal parts behind the front panel or on the subchassis section attached to it. Large, heavy components such as transformers should be centrally located, preferably near the bottom of the chassis, to keep the instrument stable. Consider heat dissipation when placing power resistors, power transistors, heat sinks, or any other device that consumes a lot of power. A large power resistor mounted near a transistor is just asking for trouble. A simple method of dissipating the heat from a power resistor is to install it without shortening its leads. Use the extra lead length to keep the resistor above the rest of the components. The location of heat sinks should be designed to minimize internal heat buildup. Heat sinks can often be mounted externally, or power transistors can be mounted externally.

Thoughtful positioning of the parts and terminal strips can eliminate excessive amounts of hookup wire. The ideal electronic circuitry would be designed so that no connecting wires would be necessary. (Printed circuits and plug-in connectors are beginning to approach that ideal.) All connections would be made with the component leads connected to other component lugs or to terminal board lugs. The ideal wiring job has yet to be achieved, but you should always work to eliminate as many wires as possible. When you look at the layout of a complicated electronics project and notice very few wires, you can be sure that a lot of thought and effort have made this apparent simplicity possible.

When you install capacitors, precision resistors, or any other component whose value is marked on its body, position the part so that its value can be read easily. It's no fun when you try to check part values and find that they're turned inward toward the circuit. Easy-to-read part values are a great help in locating and identifying circuits when you are troubleshooting.

Another factor that always has to be considered—and is frequently overlooked—is the electrical interaction of components, wires, and chassis to one another. In direct-current (dc) or low-frequency circuits this problem seldom has much effect (except for 60-Hz ac). In high-frequency circuits, however, unwanted capacitance, inductance, and magnetic and electrostatic fields are always a problem. The layout of parts in these types of circuits is critical. Many electrical interaction problems can be minimized by keeping the inputs away from outputs, keeping small signals away from large amplified signals, twisting power lines together, keeping magnetic fields at right angles to each other, and using shielded cable to minimize radiation.

Don't forget test points! Sooner or later your device is going to malfunction and need repair. A pet gripe of most service technicians is the lack of good, easy-to-reach-and-identify test points. What's a good test point? Any connection where meaningful voltage, current, or resistance can be measured. Inputs and outputs of circuits, stages, and sections are good examples.

Whenever you cover up a connection with parts or wires you have unintentionally hidden a test point. Every time a lot of components are jammed into a small area, troubleshooting becomes tougher. Identify key points with labels. Examples of such key points are B+, B−, ground, inputs and outputs, and any place where significant voltage, current, resistance, or waveforms can be measured.

Component Mounting

One of the major obstacles in electronic work is mounting the parts so that they are physically strong. Hookup wiring and small components such as resistors, capacitors, diodes, and transistors must be fastened to some type of rigid connection. Some electronic components have mechanically strong terminals that can be used for tie points. Examples are sockets, potentiometers, switches, lamp assemblies, can capacitors, and jacks. There are also many different types of terminal strips and posts that can be used for mounting parts (see Fig. 2-6). Often it is possible to mount terminal strips using existing screws. The mounting screws of a power transformer are an example. Most of them have four mounting screws that can also accommodate four terminal strips. Even the screws and nuts that hold together the case of a power transformer can be used effectively.

Fuse clips make great places to mount top-hat diodes; just snap them in like a fuse. Each fuse clip can usually hold two top hats, so if you put two clips side by side you have the means for rigidly mounting a bridge rectifier.

Mounting power transistors, SCRs (silicon-controlled rectifiers), and triacs to heat sinks can be a problem. The cases of these devices are usually connected to one of the active elements; hence they must be insulated from the sink. If the sink or device doesn't come furnished with such insulation, a mica or thermafilm insulating washer should be inserted between the case and the sink. Also, don't forget to insulate the device's mounting screws from the sink with hollow, plastic spacers. For maximum heat transfer from the device to the sink, use a thin film of silicon grease on the parallel surfaces.

Socket turrets and terminals (Fig. 2-7) are available with up to 12 terminals. Many times, the circuitry of an entire stage can be

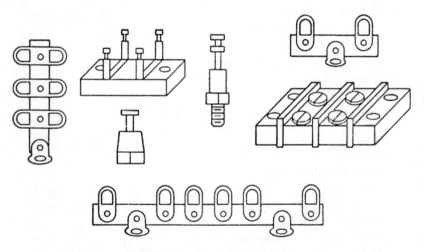

Fig. 2-6.   Common terminal connectors.

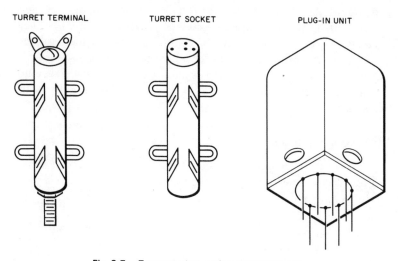

TURRET TERMINAL   **TURRET SOCKET**   **PLUG-IN UNIT**

Fig. 2-7.   Turret terminal, socket, and plug-in units.

mounted directly on one of these turrets, making a clean, compact wiring setup. Turrets are made for transistors and plug-in units. Round and square plug-in units with aluminum shield casings make excellent places to mount more components above the chassis. Such plug-ins are made to fit all standard tube sockets, which may be a stumbling block because standard tube sockets are getting a bit rare.

If you are building a circuit with a lot of common connections (e.g., a common-ground circuit), use a bus bar along the length of the circuit or along the chassis itself. If that's not too feasible, scatter a few terminal lugs around the chassis in convenient points and fasten them with screws and nuts. These are good for ground connections only, unless you use hollow plastic spacers here, too.

It is usually bad practice to mount components on top of other components. There are a few exceptions. A resistor that is fastened solidly at both ends can support a parallel bypass capacitor fastened to the resistor leads, or vice versa. Or you can mount a small radio-frequency (RF) coil in the same manner. Just be sure the resistor doesn't dissipate too much heat in its normal operation.

Terminal boards make it easy to mount a lot of parts in a row. Most terminals are swaged on the nonterminal side of the board. When you fasten the board to the chassis or whatever, remember either to insulate the bottoms of the terminals or to standoff the board. Then nothing will be shorted together. Speaking of standoffs, there are individual standoffs available, both insulated and noninsulated. They come with threaded holes for mounting them with screws, and they are great for making rigid connections where space may not permit mounting a terminal strip or a terminal lug directly on the chassis.

Wiring Techniques

The wiring techniques discussed in this section are recommended for electronic equipment that is not affected by stray wiring capacitances or inductances and that is not covered by a kit builder's manual.

Before beginning to run wires between components, study the layout of the parts and terminal strips. Try to visualize (if you haven't done so while you were designing the parts layout) a central path for all the wires to follow, branching off at various intervals to connect a series of components. Figure 2-8 shows a typical wiring layout using this technique. Notice the main bundle in the center of the layout, with branches leading horizontally to the various components. All single wires are run into groups and then into the main bundle. When it isn't practical to have one main wiring bundle, use two or three.

The secret of a clean wiring job is to make it appear as though there are very few wires connecting the parts. Keeping the wires in bundles is one way to create this effect. Another is to keep the wires close to the chassis. Always dress a wire from a connection along the chassis and into a bundle, as shown in Fig. 2-9. Whenever you are running a wire by itself, there probably is a better way to reroute the wire so that it can be run in with other wires. Keeping the wires close to the chassis and in bundles makes all the terminals and connections easily accessible for tests and troubleshooting.

Two cautions about bundling: First, use color-coded wires for yourself. For example, use red wires for B+, green wires for signal paths, black wires for grounds, and so on. Color coding makes it much easier for you to track down malfunctions, especially when you have your wires in bundles. Second, don't bundle any wires carrying 60-Hz ac with any wires carrying signals in audio equipment. The possibility of hum is too great. Wires carrying ac should be twisted

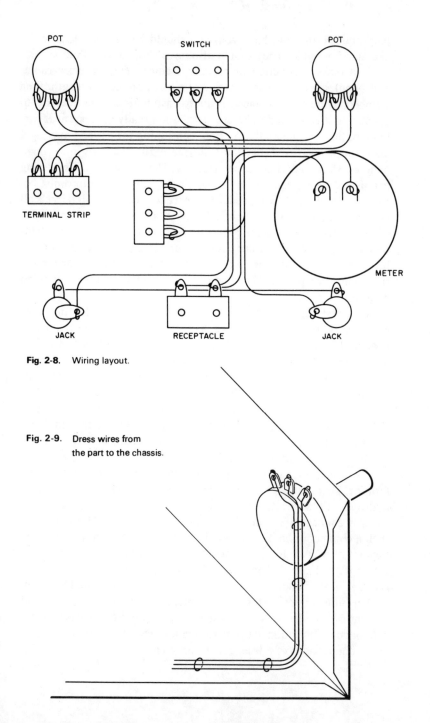

Fig. 2-8. Wiring layout.

Fig. 2-9. Dress wires from the part to the chassis.

pairs, quads, or what have you, and should be kept as close to the chassis as possible. They should also be as short as possible.

In mass production of electronic equipment, the wire cables and bundles are made separately. The wires are cut to length, then stripped, bundled, and laced prior to installation. This method of producing wiring bundles before installation really makes a neat wiring job, but it is generally too time consuming for anyone making a one-shot project. A technician will usually run the wires into rough bundles while soldering the connections. Then, after all the soldering is finished, he will lace or tie the bundles together. A compromise between both methods will produce an excellent cabling job. Here's all you do. Cut your wires slightly longer than necessary. Fasten one end of all the wires to their connections. Then run the wires into bundles. (Here, especially, color coding your wires is a must.) Now tie or lace the bundles, then fasten the remaining wire ends to their respective connection points. The bundles will be neat and uniform, with no bulges from excessive wire.

A first-class wiring job will never have any floating connections. Before it is soldered, each connection must be mechanically anchored to a terminal strip, part lug, standoff, or other firmly fastened tiepoint. One end of a resistor connected to a wire, or two or more resistors or capacitors connected in series, are examples of floating connections. Such floaters are common in inexpensive electronic equipment and are often put in by inexperienced amateurs. (Old-time pros do it quite often during repairs. They either can't get exact replacement parts or are too lazy to get them, so they string together the parts they have handy.)

Another poor wiring technique is the overloaded connection. A terminal strip, lug, or whatever, becomes a convenient tiepoint for a multitude of connections. When an overabundance of parts and wires is connected to one place, you've got a mess. (Not only does such a connection make repairs difficult, the amount of metal at the tiepoint may draw too much heat from your iron, resulting in a cold solder joint. Keep that in mind when you troubleshoot older equipment. A heavy concentration of wires and parts leads could be a possible trouble source.) Most of the time, good design and layout will prevent overloaded connections from happening. For the cases where overloaded connections appear to be inevitable, try to parallel some of the connections to a free, convenient tiepoint.

When you are wiring amplifiers or other devices that have very weak signals, keep the input and output wiring separated. Capacitive or inductive coupling can occur between parallel wires, causing undesirable feedback.

Lacing and Wire Harnessing

Lacing and wire harnessing put the final touch on an electronic project and turn an average job into a professional-looking piece of work. There are a number of methods used to cable wires together. Some lacing materials cost practically nothing; others are fairly expensive. The finished wiring will always appear neater regardless of the lacing or harnessing method used. Probably the most simple method of lacing consists of tying your bundle with some kind of cord: fishline, kite cord, dial string, and the like.

Plastic spiral wrapping (Fig. 2-10) also works well. The spiral wrapping has the added advantage of providing some mechanical protection for the bundle. Wires can enter or exit anywhere along the spiral wrap. Wire additions are fairly easy to include in bundles when plastic spiral wrapping is used.

Cable ties are another popular method for harnessing wires. Figure 2-11 shows some of the common cable ties, all of which can be fastened by hand. Many companies prefer cable ties because they can be installed quickly. For the individual project, cable ties are no better than any other type of harnessing. Just keep the wires parallel, the bundle firm, and the harnessing uniform, and you will be happy with any of the techniques discussed.

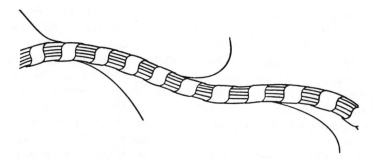

Fig. 2-10. Spiral wrapping harness.

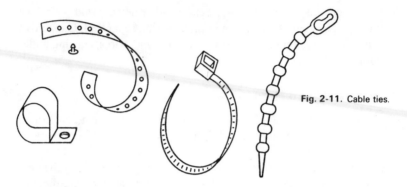

Fig. 2-11. Cable ties.

Soldering

Soldering can make or break any electronic project. If you use the proper tools and correct techniques, soldering is a breeze. Use the wrong tools or be careless, and it's disaster time. Electronic kit manufacturers claim that 90 percent of the completed kits returned for repair, or because they don't work when they are fired up, have poor connections and bad soldering. As for troubleshooting, cold solder joints are a sure way to drive a troubleshooter up the wall. Joints that look fine but are electrically open or create high-resistance points in a circuit are called "cold joints." The term *cold* means that the solder was not sufficiently melted and/or the flux was not completely removed from the joint. Solder "bridges," "icicles," and "hairs" aren't much help, either. Bridges are excess globs of solder between two points not meant to be connected; they can be caused by the manufacturer, but very often they are due to excess solder applied by a "more is better" amateur solderer. Icicles are pieces of solder that look like icicles hanging down (or standing up) from a solder joint. By themselves, they're nothing, but if they're too long, bent, or broken off, they can cause short circuits, just lke bridges. Icicles occur mostly in production soldering of PC boards, but the amateur solderer can cause them, too. Solder hairs are very fine strands of solder that look like a human hair or a strand of a spider web, but they are thick enough to create short circuits in signal paths, especially when the signals are in the millivolt and microvolt ranges. They are rare, but if you have a problem and you see what appears to be a harmless human hair, a web, or even a "dust devil," get it out of there and check the circuit again.

Required Tools

The tools required for soldering hard-wired circuits are few but important. Some are the same as those listed in Table 2. The remainder comprise the proper soldering equipment; some heat-sinking type tweezers (Fig. 2-12A) for solid-state or small-component soldering; and a braid cleaner (Fig. 2-12B) for cleaning component leads before their insertion.

Determining the proper soldering tool for the job depends on many things. For example, if you are going to be soldering for any length of time, it pays to use an iron: it's always "on." On the other hand, if you're going to do one or two quick connects or disconnects, a gun is preferable: warmup time is generally shorter than that for most irons. Then again, there might be the case where both tools are required: you are going to solder for a long period of time, but in some cases may need the extra lighting furnished by a good gun. Also, a gun is somewhat like the "hatchet" iron discussed later, and may be easier for you to use than an iron. Whichever tool you use, gun or iron, the following information will help you, even though it is based on the use of an iron.

Selecting the proper iron for the job depends on many things: wattage, weight, length, balance, hand-to-tip proximity, and color are the considerations that combine to produce the one proper soldering instrument for the job to be done. Soldering hard-wired circuits is the kind of work in which the iron may idle for prolonged periods, while you do some more wiring tasks, or troubleshooting, or whatever. Therefore, the iron should not overheat, but still should be readily available for soldering without tip retinning or dewetting problems. Further, when it comes to irons, wattage is one of the greatest misnomers you can encounter. For years, soldering irons have been selected on the basis of wattage alone. The really important factors to be considered are:

- Tip mass—large mass, high reserve
- Tip length—the shorter the tip the more efficient the iron
- Frontal mass of iron—the greater the mass, the lower the iron's idling temperature.

Ideally, it is best to use the smallest, shortest, and lightest iron with the necessary thermal capacity. Excessive weight promotes premature fatigue. Excessive length will cause the iron to be un-

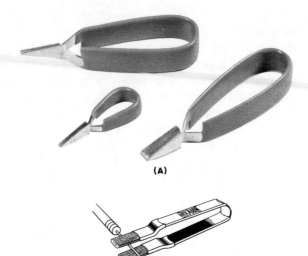

(A)

(B)

Fig. 2-12. Soldering accessories. (A) Heat-sinking tweezers. (*Courtesy* Hunter Tools) (B) Component-lead cleaner. (*Courtesy* Hexacon Electric Co.)

Fig. 2-13. Hatchet-type iron. (*Courtesy* Hexacon Electric Co.)

balanced, with the hand too far from the tip for adequate directional control. Thought should be given to the advantages of "Hatchet"-type soldering irons, such as that shown in Fig. 2-13. They are usually very well balanced; the hand is close to the tip; your arm is in a natural position while holding the iron; and they minimize handle-heat problems. Handle color has also taken on new importance;

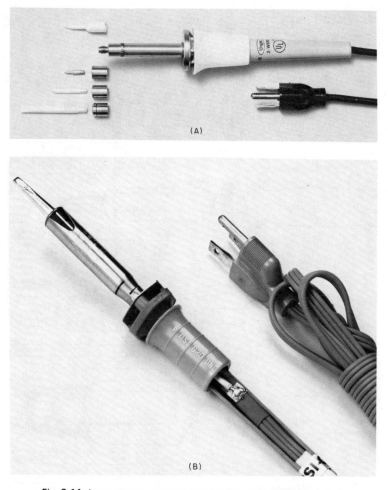

Fig. 2-14. Low-wattage, grounded-tip soldering irons. (A) (*Courtesy* Ungar Div., Eldon Industries). (B) (*Courtesy* Hexacon Electric Co.)

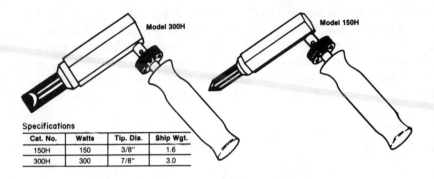

Specifications

Cat. No.	Watts	Tip. Dia.	Ship Wgt.
150H	150	3/8″	1.6
300H	300	7/8″	3.0

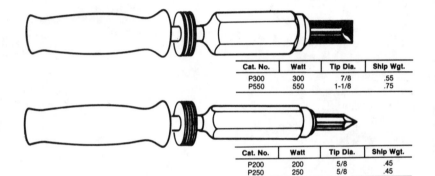

Cat. No.	Watt	Tip Dia.	Ship Wgt.
P300	300	7/8	.55
P550	550	1-1/8	.75

Cat. No.	Watt	Tip Dia.	Ship Wgt.
P200	200	5/8	.45
P250	250	5/8	.45

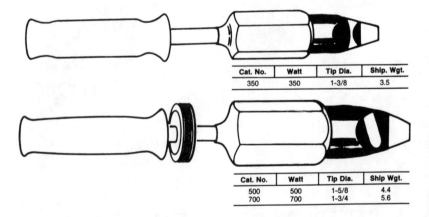

Cat. No.	Watt	Tip Dia.	Ship Wgt.
350	350	1-3/8	3.5

Cat. No.	Watt	Tip Dia.	Ship Wgt.
500	500	1-5/8	4.4
700	700	1-3/4	5.6

Fig. 2-15. Heavy-duty soldering irons. (*Courtesy* Hexacon Electric Co.)

studies have proved that certain colors are more visible at the edge of your peripheral vision. Proper handle colors promote safety and efficiency, and lessen your chances of grabbing the hot element instead of the handle (unless, of course, you have the iron in an enclosed stand). Further, the handle should have a large enough flare at the element end to prevent your hand from sliding onto the hot metal element. What it comes down to, then, is that for hard-wire soldering you need at least a grounded-tip soldering iron, such as those in Fig. 2-14, for standard soldering; and a soldering iron or gun (Figs. 1-11, 1-12, and 2-15) with higher thermal capacities for soldering to chassis.

For the soldering iron(s) you should have a stand, like those pictured in Fig. 2-16, or the more exotic one shown in Fig. 1-8B. On the other hand, if you want to use your money for other purposes, you can easily make your own holders. Just get hold of some old typewriter ribbon reels and flatten their rims, as illustrated in Fig. 2-17. The flattened edges will support the reels on a workbench, and the irons' tips or bodies can then rest on the reels' axles. Flare the top of one reel slightly, to accommodate the large iron.

An exceptionally fine tool that can be purchased is the SOLDERHAND (Fig. 2-18). This new tool clamps onto any pencil type of soldering iron and lets you feed the solder and do the soldering with one hand. This frees your other hand to hold or manipulate the piece being worked, or to do any other soldering task normally requiring three hands. Also buy some soldering aids, such as those shown in Fig. 2-19. They're great for probing, holding down wires while soldering, wire brushing just-finished and old solder joints, and so on. Last, buy some tweezers, such as those shown in Fig. 2-20, for odd jobs such as picking up a tiny diode that slips from the connection before it's soldered, or holding small components while you tin the leads.

Soldering Decisions

What should you use? A soldering iron or a soldering gun? The answer is simple. If you are making only a few solder joints, as in repair work, your best bet is the fast-heating gun. On the other hand, if you are going to be soldering for hours or days, as in construction work, then use the iron. Even the few seconds it takes for a gun to heat is too long to wait when you have a lot of soldering to do.

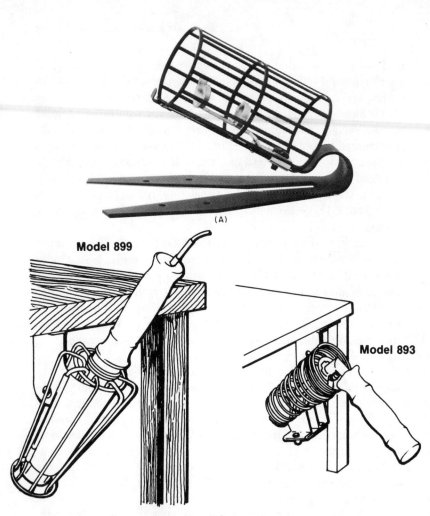

Model 899

Model 893

(A)

(B)

Fig. 2-16. (A) Soldering iron stands. *(Courtesy* Ungar Div., Eldon Industries) (B) Models 893 and 899 adjustable, under-bench-mounted soldering iron stands. *(Courtesy* Hexacon Electric Co.)

Fig. 2-17. Shaping ribbon reel to serve as iron holder.

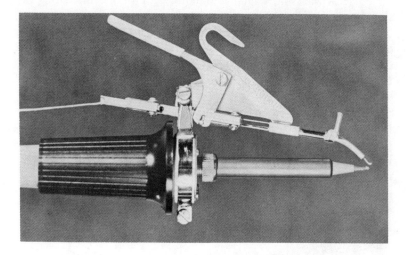

Fig. 2-18. SOLDERHAND soldering tool attached to a pencil type of iron. *(Courtesy* Lintek Co.)

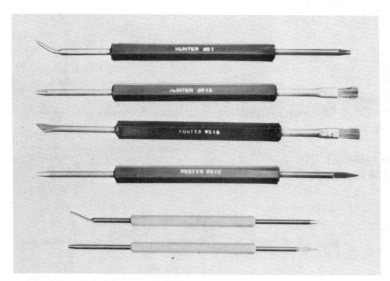

Fig. 2-19. Soldering aids. *(Courtesy* Hunter Tools)

The perfect solder joint is smooth, with a uniform coating of shining solder. The resin core flux has been boiled away, and all the connections on the joint can be seen but appear to be one. If you apply too much heat, the wire insulation will melt back (unless you use the anti-wicking tweezers of Fig. 2-20A) or parts will be overheated. Use too little heat, and you have a cold, crystalline, grainy solder job.

The wattage written on the gun or iron is one clue for proper selection. If you are trying to solder a ground terminal connected to five or six wires or parts, use a high-wattage gun or iron. A gun or iron of 250 watts may be necessary if the terminal is fastened directly to a large, heavy chassis. The chassis acts like the heat sink, and unless you have enough heat it will absorb the heat as fast as you apply it. If there are parts tied to the terminal that could be damaged by ex-

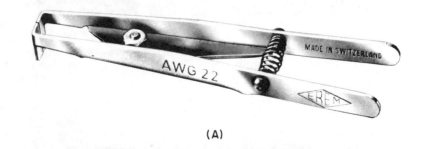

(A)

(B)

Fig. 2-20. Tweezers. (A) Nonwicking; (B) general-purpose, precision pointed. *(Courtesy* EREM Corp.)

cessive heat, clip a heat sink between the terminal and the part (Fig. 2-21). An ordinary alligator clip can also make a good heat sink.

Most hard-wiring connections can be made with a 15-47 watt soldering iron. The secret for good results from any iron or gun is to get good heat transfer from the heating element to the soldering tip. Many soldering irons have tips that screw into the heating element similar to those seen in Fig. 2-14A. The tip threads oxidize eventually and bond to the element unless periodic care is taken. Generally, all that is needed to break this resistance is to loosen and tighten the tip a few times. Some soldering iron manufacturers recommend an anti-seize compound (e.g., Ungar #8001 Anti-Seize), which helps to prevent the tip from becoming fused to the element. After every 20 to 30 hours of iron use, remove the tip and reapply some anti-seize compound to the threads. The tips of soldering guns will often oxidize where they are fastened to the gun, causing loss of heat transfer. The gun body heats up instead of the tip. Loosening and tightening

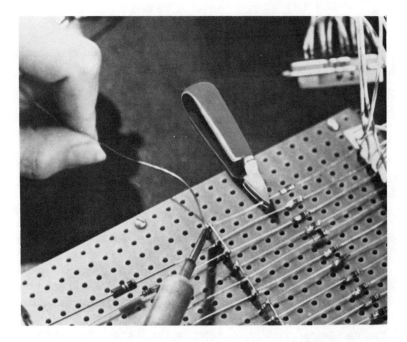

Fig. 2-21. Using the heat sink to avoid thermal damage. *(Courtesy* Hunter Tools)

the connecting screws or nuts a few times will serve to eliminate this problem.

Keeping the soldering tip clean and tinned is very important to good soldering. A damp sponge, rag, or paper towel makes a good wiper for soldering tips. Copper tips should be filed to shape when they become pitted. Before you file the tip, make sure it isn't an iron-clad type: filing removes the thin iron plating. Also, on an ironclad tip use only a damp sponge (e.g., Ungar #400 Kleen-Tip Sponge and Tray, which can be affixed to your bench) to clean the tip; the iron plating is thin enough to be rubbed away by cloth or paper, even if the cloth or paper is wet.

Follow these soldering pointers for professional results.

- Make sure the connections to be soldered are physically tight and clean.
- Tin the cleaned soldering tip with whatever type of solder you are using, preferably 60/40 tin-lead, resin-core solder for electronics work.
- Position the soldering tip, if possible, at the bottom of the joint so that gravity will help the solder flow.
- Keep as much contact as possible between the soldering tip and the connection surface for quickest and maximum heat transfer.
- Apply solder to the top of the joint if possible. The solder should melt almost immediately and flow downward over the joint toward the soldering tip. Use just enough solder to wet the connection thoroughly.
- Remove the tip and solder when you see the flux start to bubble, making sure the joint doesn't move until the solder has hardened. If you want to speed up the solidifying time of the solder, touch the joint with your wet sponge.

Always use care during your soldering; avoid the possibility of creating uncalled-for problems. When carelessness creeps into the picture, soldering can become very expensive. For instance, a splash of solder gone unnoticed on a peaking coil (Fig. 2-22) can upset the entire performance of your project. If the splashed solder is hidden, it may even lead into a perplexing and time-consuming analysis of malfunctioning circuitry. A short in the B+ circuits caused by splashed solder can result in excessive damage to your project or in a frustrating case of intermittent operation.

SOLDER
SPLASHED ON
PEAKING COIL

CAN CAUSE
PHASE SHIFT
AND AFFECT
FREQ. RESPONSE

Fig. 2-22. Solder splashes can
affect performance.

When buying a soldering iron, remember that the wattage and size are not the only considerations. The iron should also be well balanced and easy to hold. Furthermore, the handle should remain cool and comfortable enough to hold even if the iron has been on for hours. Check to make sure that the element end of the handle has no metal fastened to it that can burn your fingers.

Although more and more electronic equipment is being made with printed circuits and modules, there are still a lot of conventional hard-wired circuits around. Even equipment containing printed circuits may have components that are hard-wired to a front panel or to the chassis. One of the more difficult of these hard-wired jobs is the multiple connection with many parts and/or wires connected together. To make soldering this type of joint easier, proceed as follows: Start the connection at the lowest point on the terminal strip, lug, or whatever and fasten it securely with a mechanical connection. A single bend of the wire squeezed tightly to the lug will do. Connect the next wire or part directly above the first, again with a mechanical connection. Continue in this fashion until all your connections are made, then solder them all at once. You'll probably need a higher wattage iron or a gun to do it. The finished joint will be a thing of beauty and, just as important, easier to desolder and disconnect when trouble crops up.

Last, even though you have a beautiful shiny joint, it might just be a cold solder joint. Only experienced technicians can tell simply by looking. So, to avoid problems, take an ohmmeter and check your just-made connections for continuity. First, gently wiggle the connection points; this will probably ensure that any possible breaks will occur. Now, with the meter set at the R × 1 scale, apply the meter leads to each end of the connection but *not* to the joints themselves. You want to make sure that the *components* are connected, not just the solder holding them tight. This is a tedious procedure, true. However, once you are finished, you can be a lot more sure that it's not your soldering that's at fault should your project not work then and there, or any time in the future.

Wire Wrapping

As the electronics industry has advanced in technology, the need for a faster, more reliable, and less expensive method of making electrical connections during manufacture has become a necessity. In past years it was enough to solder most connections because, by today's standards, there was plenty of room between terminals and relatively few connections. However, modern electronic equipment, especially computer mainframes and patchboards, is far more complex, with many more terminals. The task is compounded by the ever-increasing demand for reduction in size. The end result is far more terminals in much less space. To solve this problem, the industry has had to adapt to a completely different type of connection. The solderless connection achieved by the wire-wrapping technique is now a standard method of making terminal connections in high-density electronic equipment. Further, wire wrapping may be catching on with those electronics hobbyists and experimenters bent on putting together their own microcircuits. With wire wrap you don't need artwork for PC boards: you can obtain commercial wire-wrap plug boards, circuit card connectors, assorted wrap posts, wire dispensers, DIP wire-wrap sockets for ICs, numeric LED displays, and DIP switches. However, it's not all gravy with home wire wrapping: your projects cannot be micro in physical size; for example, you can't build a pocket calculator using wire-wrap techniques. Also, for one particular type of wire wrap, "pencil wire," you still need the good old soldering iron or gun to make a proper connection. Lastly, if you use too heavy a wire gauge you can put so much strain on the wrapping posts that they can break or pull out of or break the plugboard.

Figure 2-23 illustrates the four types of wire-wrapping tools that are available. They are the pneumatic or air type, used mostly

ELECTRIC   BATTERY   AIR   MANUAL (Chuck-Type)

Fig. 2-23. Wire-wrapping tools. *(Courtesy* OK Machine & Tool Corp.)

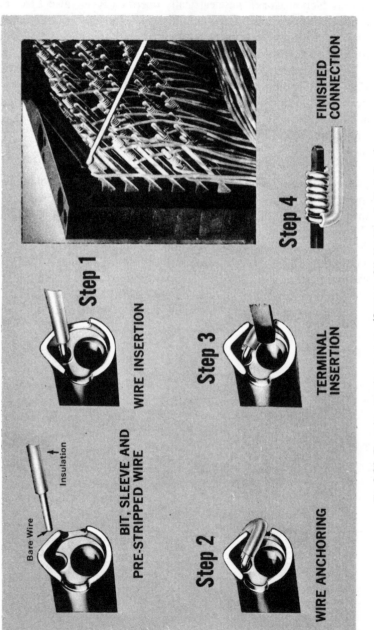

Fig. 2-24. The wire-wrapping process. *(Courtesy OK Machine & Tool Corp.)*

for production work; the electrically operated types, also used for production work; and battery-operated or manually operated types, used for prototype, service, and repair work. Refer back to Fig. 1-7 for another manual type of tool.

The wire-wrapping process itself is fairly simple, as shown in Fig. 2-24. First, the wire (solid wire only, not stranded) is bared sufficiently to permit several turns (generally, at least six) around the applicable terminal; charts are available to determine this length. The bared wire is then inserted into the smaller hole in the tool bit, then anchored by wrapping it around the notch in the sleeve. (See Fig. 2-25 for illustrations of the bit and the sleeve.) Next, the wrapping tool is inserted over the terminal, with the terminal going into the larger bit hole. The final step, the wrap, is then accomplished either by squeezing the tool's trigger (Fig. 1-6) or by twisting manually (Fig. 1-7). The end result—a regular wrap or a modified wrap (Fig. 2-26)—depends on the user's needs and the bit used. The regular bit wraps only the bared wire around the terminal. The modified bit wraps a portion of the insulation around the terminal, in addition to the bared wire, which enhances the ability of the wrap to withstand terminal vibration. Figure 2-27 illustrates terminals suitable for

BIT AND SLEEVE

Fig. 2-25. Wire-wrapping tool bit and sleeve. *(Courtesy* OK Machine & Tool Corp.)

REGULAR WRAP

MODIFIED WRAP

Fig. 2-26. Regular and modified wire wraps. *(Courtesy* OK Machine & Tool Corp.)

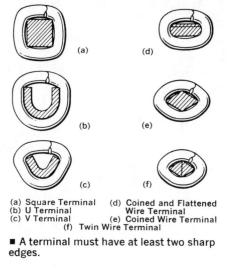

(a) Square Terminal   (d) Coined and Flattened
(b) U Terminal   Wire Terminal
(c) V Terminal   (e) Coined Wire Terminal
           (f) Twin Wire Terminal

■ A terminal must have at least two sharp edges.

Fig. 2-27. Terminals suitable for wire-wrapping connections. *(Courtesy* OK Machine & Tool Corp.*)*

making wire-wrapping connections. Figure 2-28 offers some hints on making the connections.

Printed Circuits

If you are putting together some electronics project—say, a microprocessor or minicomputer—chances are that you've bought a kit with all the components and an instruction manual included. Barring this, you may have decided to build something you've seen in a magazine such as *Popular Electronics,* or one of the projects in the Hayden projects books which supply a schematic diagram, a parts list, and a full-size reproduction of the foil side of the PC board you will need. On the other hand, there's the off chance that you want to do everything yourself from scratch. Whichever method you select, what you read here will be helpful.

Base Materials

Rigid PC base materials are available in thicknesses ranging from $\frac{1}{64}$ to $\frac{1}{2}$ inch. The important properties of the usual materials are given in Table 3. The most widely used base material is the NEMA-XXXP paper-base phenolic.

Table 3. Properties of Typical PC Board Base Materials

Material	Punchability/ Drillability	Mechanical Strength	Moisture Resistance	Insulation	Arc Resistance	Max. Temp. (°F)
NEMA-XXXP paper-base phenolic	good	good	good	good	poor	220
NEMA-XXXPC paper-base phenolic	very good	good	very good	good	poor	220
NEMA-FR-2 paper base phenolic, flame-resistant	very good	good	very good	good	poor	220
NEMA-FR-3 paper base epoxy, flame-resistant	very good	very good	very good	very good	good	220
NEMA-FR-4 glass-fabric base epoxy, general-purpose, flame-resistant	fair	excellent	excellent	excellent	very good	265
NEMA-FR-5 glass-fabric base epoxy, temperature and flame resistant	fair	excellent	excellent	excellent	very good	310
NEMA-G-10 glass-fabric base epoxy, general-purpose	fair	excellent	excellent	excellent	very good	265
NEMA-G-11 glass-fabric base epoxy, temperature-resistant	poor	excellent	excellent	excellent	very good	310

EASY DOES IT!

Don't press too hard. Let the OK tools do the work. Excessive pressure can lead to overwrapping.

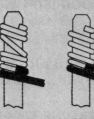

OVERWRAP

STAY WITH IT!

Just keep the OK tool on the terminal until the wrap is complete. Early removal can result in spiral and open wraps.

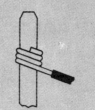

OPEN WRAP   SPIRAL WRAP

FEED WIRE CORRECTLY!

It's easy to feed wire into the slot in the OK bit correctly. Be sure the stripped end of the wire is "pushed-in" all-the-way.

INSUFFICIENT TURNS

USE THE CORRECT BIT AND SLEEVE!

Wire-wrapping is a precision technique and the wrong bit and sleeve just can not do the job. So check the OK specification tables and use the right bit and sleeve. Improper selection can cause problems ranging from "Pigtails" to loose wraps.

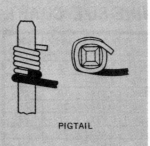

PIGTAIL

Fig. 2-28. Wire-wrapping hints. *(Courtesy* OK Machine & Tool Corp.)

Conductor Materials

Copper is used almost exclusively as the conductor material, although silver, brass, and aluminum can be and have been used. The most common thicknesses of the foil are 0.0014 and 0.0028 inch, and widths range anywhere from 0.001 up through 0.4 inch.

Design Considerations

Before you make your PC layout, you should breadboard and test the circuit under conditions as close as possible to your final operating conditions. (This applies only to your own design; kits and magazine articles will have already done this for you.) This will permit you to detect and correct any operating deficiencies and quirks before you waste your time making up the board itself. Remember that some designs may operate differently on a PC board than on a breadboard, and appropriate corrective actions may be necessary. For example, inductive coupling between foil patterns may cause undesired oscillation in high-frequency or amplifier circuits.

Hole drilling should be done very carefully to avoid chipping or separating the base material layers. As for the hole sizes themselves, use a drill not more than 0.020 inch greater than the diameter of the wire or component lead to be inserted. The terminal area itself should have a diameter that is (1) 0.020 inch larger than the diameter of the flange or projection of the flange on eyelets or standoff terminals, or the diameter of a hole that you are going to plate through; and (2) 0.040 inch larger than any hole that will be unsupported.

Conductor widths must be adequate to handle the currents you expect. Generally, a width of 0.060 inch is acceptable. As for spacing between conductors, there must be at least 0.025 inch between conductors having voltage differences of 0-100 volts, 0.050 inch from 100-250 volts, and 0.10 inch from 250-500 volts.

Artwork

The schematic diagram for your project can be obtained in several ways. You may have seen a schematic diagram of some device—say, an automobile tachometer—and have decided to use it as the basis for your PC board. Or you may have picked up both the schematic diagram *and* the actual size foil diagram, suitable for 1:1 reproduction, and decided to use those. Finally, you may have decided to design your own circuit or redraw an existing circuit, adding to it or simplifying it.

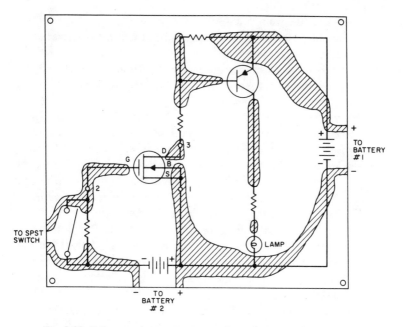

Fig. 2-29. Foil pattern layout traced over schematic diagram.

If you are using an existing diagram or your own schematic diagram, you can use the diagram itself to make your layout (i.e., if it isn't too large). For example, Fig. 2-29 shows the schematic diagram of a simple go/no-go test circuit for metal oxide semiconductor (MOS) transistors superimposed on a foil pattern that would let you go directly from a schematic diagram to a PC board. When you do such PC outlining, keep the following points in mind:

- Determine where the external connections (input, output, power supply, etc.) are going to be made. Are you going to have wires connected to one end of the board, both ends, throughout the board, or what?
- Try to position the foil pattern so that no extra connecting wires are needed on top of the board or on the foil side. The ideal printed circuit has no conductors other than the foil and the component leads.
- Keep in mind the mounting of the PC board. If the board is going to be fastened directly, provide foil areas for fasteners if the

foil is to be grounded. (A ground or common path around the board's perimeter is recommended, but it is not absolutely necessary.)

- Try to leave as much foil area as possible for parts that run hot. Some of the heat can be dissipated through the copper foil.
- You can use tracing paper to lay out the foil, but it is too flexible and may absorb too much ink, if that's what you are going to use. Try to use Mylar or Cronar base drafting material or even the old linen-base material that's still around. As for the tracing itself, you can use ink (preferably India ink) or pencil, as long as it's legible. Or you can use preformed, self-adhesive layout patterns that are available commercially.
- Never use a sharp corner when laying out your foil pattern. Always round the edges of such corners, including the inside edges. Sharp corners set up resistance pockets, which may change your circuit values to say nothing of generating unnecessary heat.
- Draw the centers of your drill holes so that their final size is no smaller than $\frac{1}{32}$ inch. (The reason for saying *final* size is that you may have decided to go oversize and then have reduced your foil pattern photographically.)

Once the rough foil pattern is established, draw another pattern in the exact size. How small do you want the board to be? If you want the board to be as small as possible, component size will determine its overall size. Do you want to mount the resistors and capacitors lengthwise or vertically? Will the transistors and/or integrated circuits be soldered directly into the circuit, or are they going to be placed into sockets? Is the board itself going to be plugged into a PC strip socket, or are the external connections going to be hard wired? Decide the answers to these questions and then design your project.

After the foil pattern is finished in the final size, lay the parts in their positions to make certain enough room is available for each one. Don't be disappointed if the first try on the pattern is a failure. It may take three or four practice versions before a satisfactory design is established.

Making the Printed Circuit

After the final foil pattern has been designed, it must be transferred to the PC board. A simple way to do this is to place a piece of carbon paper, carbon side *down,* on the blank copper and put the foil pattern on top of the carbon paper. Make sure you hold everything down tightly, and then trace the pattern with a No. 2½ or No. 3 pencil. Check continuously to make sure the carbon is transferring the layout to the copper. Copy all the foil outlines and all the connection dots.

When the transfer has been completed, lightly center punch all the hole positions. Then apply resist paint or ink to the foil paths with a small artist's brush. If some of the resist runs out of the foil paths, don't worry about it until it dries. Then just scrape the excess resist away with a razor blade. This is a lot less messy than trying to correct the error while the resist is wet.

In about 30 minutes the resist should be dry enough to permit placing the board in the etchant solution. Most etchant solutions will remove the copper foil in about 20 minutes. The etchant can be used again and again, but it will take a little longer to remove the copper each time. Use a glass or plastic tray to hold the PC board during etching. When all the copper has been removed, wash the PC board in water and inspect it carefully. If you're satisfied with the job, you're ready to remove the resist. Almost any solvent will remove the resist: turpentine, mineral spirits, kerosene, or the like. After the resist is dissolved, drill the mounting holes in the PC board. Then clean the foil with fine steel wool, wash the board again to remove any steel wool leavings, and you are ready to mount the components.

This is one simple, easy method for making a printed circuit. There are many others. You can use pressure-adhesive resist tape and circles, for example. Rolls of resist tape are available in many widths: ⅟₃₂, ⅟₁₆, ³⁄₃₂, ¼ inch and others. Resist circles are also available in various diameters. After the circuit pattern has been established, stick a resist circle at each hole position and connect each circle with the tape to produce the various foil paths. When the circuit is finished, etch the board, remove the tape and circles, and you will have a professional-looking printed circuit. Figure 2-30 shows the layout of Fig. 2-29 using resist tape and circles. Both printed circuits will work fine.

Here's another successful technique for making printed circuits, using a photographic process. Draw the foil pattern (on Mylar) larger than full scale. Use India ink (preferably) or a black felt-tip pen to black out the various areas of the circuit. (To make sure that you haven't overlooked anything or left pinholes in the inking, hold

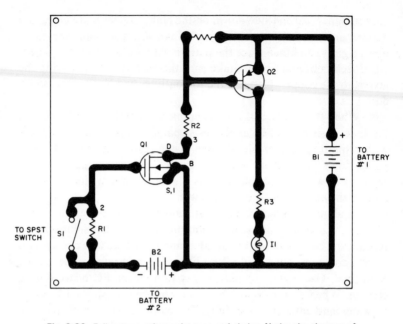

Fig. 2-30. Foil pattern using resist tape and circles. Notice the absence of square edges. Although it is difficult with tape, square edges can be removed with an X-Acto knife or a razor blade. Square edges set up high-resistance points in the circuit.

the pattern up to the light. You'll be surprised at how many seemingly dense inked areas are thin and "holey.") If you black out the foil areas that are going to be etched, you can use the photo negative. If you black out the foil areas that are going to be saved, you must use the positive print. This positive print must also be translucent.

If your original schematic diagram has reference symbol numbers on it, copy these numbers onto your layout in the places where your components will be mounted, as shown in Fig. 2-30. If your schematic diagram has no such numbers, put in some. Such identification facilitates both initial mounting and later troubleshooting. It's also a good idea to include polarity markings and pin numbers of solid-state devices.

By making the original pattern larger than full scale, a more accurate drawing can be made. Also, small imperfections will tend to disappear; they will be condensed in the reduction process. The type of camera available will determine how large the original pattern can be made. Furthermore, if you have your choice, make the original an even-number oversize, for example, 2:1, 4:1, and so on. This makes it much easier to determine what your connector widths and hole sizes will be in final size.

As soon as the negative (positive) is finished, you are ready for the next step. Photo-sensitive PC boards are available. They are packaged in a light-proof wrapper that should be opened only in a dark room. Lay the negative (positive) over the photo-sensitive side of the PC board and expose the board to light for 10 to 12 minutes. You can use any ordinary lamp, but a good way is to position one of those flexible desk lamps, with a 100- to 150-watt bulb in it, directly over the negative (positive) and board combination. Then place the board in a developing solution (available commercially) for about 1 minute. The areas that received the light will retain a photo-resistant coating. Next, etch the board as explained previously. The photo-resist can then be removed from the foil by a piece of steel wool (an SOS pad, or equivalent, is excellent). Wash the board, and you're ready to go.

Silk screening is another method for producing PC boards. However, its usefulness lies mostly in making many identical circuits, hence it is not discussed here. (You can also use the film negative/positive to make many duplicate circuits, but making the negative/positive is a lot easier than making a silk screen.)

Another way to make your PC board is to use pre-etched, pressure-sensitive, copper-circuit patterns that are available commercially. If you choose this method, all you need is a standard laminated board without any copper foil on it and a package of these universal circuit patterns. It's easy. Just follow the schematic diagram and select the copper patterns you need. Each pattern has a paper backing. Peel off the backing and stick the pattern in place. Connect the patterns with pressure-sensitive conductor paths or even with hard-wiring. Do you want to use the board as a plug-in module? Use pressure-sensitive connector strips. If you want to reposition the patterns or use them somewhere else, just peel them off and restick them in the new place.

Assembly

With the PC board etched and drilled, it's time to mount the components. Generally, the fastest way to complete this phase of the project is to insert all the components into their respective PC board positions on the nonfoil side before soldering. Mount small resistors and capacitors so that their surfaces touch the board. However, it's a good idea to mount high-wattage parts up and away from the board so that their heat can be dissipated more readily. You can use small pieces of sleeving as spacers. If you are mounting glass diodes or other fragile parts, bend a loop in the leads before inserting them into the holes, as shown in Fig. 2-31. This loop will act as a strain relief in case the board is stressed. It will also act as a small heat sink during the soldering (and possible desoldering) process, and as a spacer to keep the part off the board for convenient test points.

If you are inserting an integrated circuit (IC), it is very important during lead bending that the leads be supported and clamped between the bend and the seal at the IC casing. Otherwise, you may break the seal or damage the lead plating. Long-nose or duckbill pliers can be used to hold the lead as shown in Fig. 2-32. In no case should the radius of the bend be less than the lead diameter or, in the case of rectangular leads, less than the thickness of the lead. It is also important that the ends of the bent leads be perfectly straight and parallel to assure proper insertion into the holes in the PC board or into the socket in the board.

One other note. If you're handling integrated circuits or metallic oxide semiconductor field-effect transistors (MOSFETs), be sure

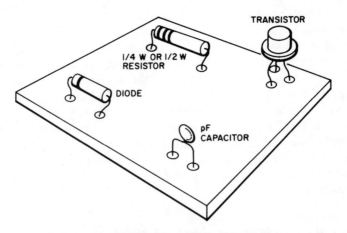

Fig. 2-31. Mounting fragile components.

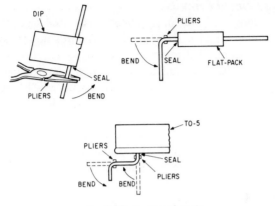

Fig. 2-32. Bending IC leads.

the hand holding the part is *grounded,* for example, through a commercially available metallic wristband.

If possible, use sockets when you are installing transistors or other solid-state components. Solid-state parts are extremely reliable and seldom go bad. Therefore, many are soldered directly into the circuit. However, when a defect does occur, it's a big help to be able to remove solid-state parts with no desoldering damage. They can be checked easily out of the circuit, and their removal provides an efficient way to isolate various electronic stages from one another.

When mounting parts, don't forget to keep the parts' value labels on the top so they can be read after installation. (It would also have been a good idea to put values in during your foil-making process. It's very helpful during assembly, and it's so easy to do when you use the photo negative/positive process.) Make sure that polarity markings are visible on diodes, rectifiers, capacitors, and other polarized devices.

As you insert the parts into the board, bend the leads flat against the board, preferably along their respective copper strips, so that the parts won't fall out before you can solder them. Paralleling the leads with the strips lessens the possibility of solder bridges and increases the probability of a good solder connection. After bending, cut off the excess of each lead to avoid later short circuits. Figure 2-32 illustrates the proper way to bend IC leads for insertion into the board.

Soldering Printed Circuits

Everything said previously about hard-wired circuits is applicable to printed circuits. Remember, PC boards don't stand alone. They must be connected to another board, to a transformer, or to something else discrete, and the connection requires hard wiring. There is, however, one difference: Microcircuit and PC boards are much more delicate than hard-wired circuits. Hence, greater care and irons of lower wattage are necessary.

Soldering in printed circuits is, in some ways, easier than hard-wired circuit soldering. First, the connections are almost always single; one lead fits into one PC hole. Second, the parts are generally not in the way when you are soldering—they're on the nonfoil side of the board. And, finally, PC connections are easier to clean for soldering. The flat copper ribbons are easy to scrape clean with a small wire-brush soldering aid (Fig. 2-19) or steel wool (only if there are no parts mounted on the foil side).

For microelectronic PCs, soldering is not quite so easy. Temperature of the tip, tip size, and tip dwell time (the length of time the tip is kept on the joint) are all critical. According to the data sheet of one manufacturer of microelectronic equipment, if the component lead at a point $\frac{1}{16}$ in. from the body of the component is subjected to a temperature of 509°F (265°C) for a period of 10 seconds, it will cause an internal failure. If the lead is subjected to a higher temperature, the corresponding time span to failure is decreased. Therefore, if you have a soldering iron that operates at a temperature substantially higher than 509°F, and since microelectronic leads and pads are so small, the lead at a point $\frac{1}{16}$ in. from the component body reaches the tip temperature almost instantaneously. Now, in order to get good alloying when soldering, the parts to be joined must be brought up to approximately 100° – 125°F (38° – 52°C) over the melting point of the solder being used. (This approaches the destruct temperature of many sensitive components; for example, ICs.) It is generally agreed throughout industry that this should be done within a dwell time of $1\frac{1}{4}$ seconds, to preclude the possibility of damage not only to the sensitive components but also to circuit boards, flex circuitry, adjacent components, wire insulation, and so on. If the soldering iron tip temperature is too low, this temperature elevation cannot be accomplished within the specified dwell time. If the tip temperature is too high, the risk of damage is greatly increased. This leads to the conclusion that you cannot be blasé about the soldering irons you use in putting together your microprocessor or other microelectronic device. You must have exactly the right combination of soldering iron and tip to derive optimum performance and safety. Thus, fixed (Fig. 2-33) or variable-temperature (Fig. 2-34) irons of low wattage and tiny tips are a must for printed and integrated circuits. For example, if your project is a microprocessor, your best bet is an iron of only 4.5-6 watts. If you are working on a more or less standard PC board, you should use, say, an iron of 10-20 watts, like those shown in Fig. 2-33. If the board, however, is going to take one or more integrated circuits, then use a 4.5- to 6-watt iron for these components. An iron or gun of higher wattage may generate enough heat to destroy the bond between the foil and the board, melt the foil, destroy the part even if you've heat-sinked it, and/or char the board itself, weakening the entire circuit.

One last item on microcircuits: ALWAYS use a 3-wire, grounded iron. And be sure the ground connection (connection between the round, third prong on the ac plug and the iron's tip) is a good one; i.e., reads continuity on your ohmmeter's RXI scale. The

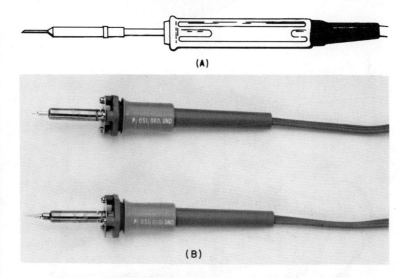

(A)

(B)

Fig. 2-33. Fixed small wattage irons. (A) Antex Model C 15-watt soldering iron with No. 6-NS tip. *(Courtesy* Antex Div., M.M. Newman Corp.) (B) Models H10/H11-Top-and H14 irons with J901X tip-top-and J905X tip; usable with "Dial-a-Temp" unit. (*Courtesy* Hexacon Electric Co.)

same oxidation that occurs with screw-in tips and prevents them from attaining full thermal capacity can also cause a high resistance ground and a low voltage appearing at the tip. Should that occur, you can blow a CMOS or MOSFET IC just like that!

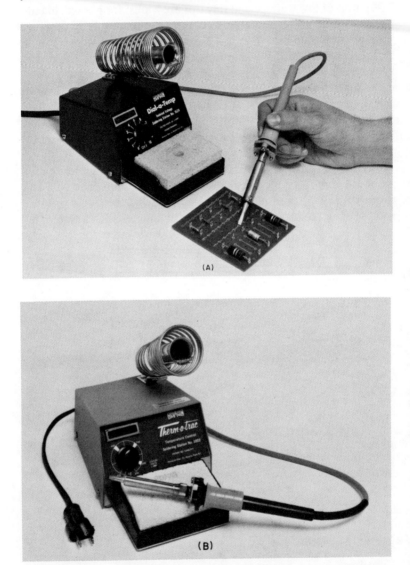

(A)

(B)

Fig. 2-34. Soldering-iron control units. (A) Dial-a-Temp unit; (B) Therm-o-Trac unit. (*Courtesy* Hexacon Electric Co.)

Position the soldering iron tip so that maximum surface contact is made between the tip, the foil, and the component lead. For fast and good soldering, it's important that the *tinned* soldering tip touch the foil and component lead simultaneously. Add your solder, and the connection should be made almost instantaneously.

Many of these low-wattage irons have interchangeable tips for various kinds of fine work. Tips that are only 1/64 inch in diameter are available. The shape and size of the tip are also factors to consider when soldering. There are straight tips, offset tips, chisel tips, pencil tips, pyramid tips, screwdriver tips, needle tips, spade tips, and slide-on and screw-on tips. Figs. 2-35 and 2-36 illustrate the various kinds of tips available.

Fine or extra fine gauge solder will generally do the best job in micro and standard printed circuits. Solder with 60% tin and 40% lead, 20 gauge, is the best all-purpose solder. If you are going to solder a lot of microcircuit or IC connections, use a solder of smaller gauge (Refer to Table 1).

At the risk of being redundant, remember that microcircuits, integrated circuits, diodes, and transistors are sensitive to heat. So, while you're working on them, be sure to heat sink their leads. And don't keep the iron on the connection until you see the flux bubble. Pull the iron away as soon as the solder melts. Remember the 1¼-second limit. If you want to make sure you don't exceed this limit,

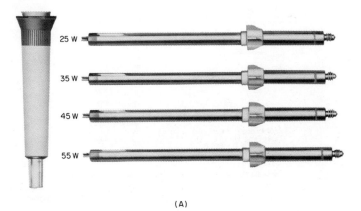

(A)

Fig. 2-35. Soldering tip shapes and sizes. (A) Imperial iron with available heating elements. (*Courtesy* Ungar Div., Eldon Industries)

Imperial Thread-On Soldering Tips

#6304 Micro
Chisel. .500
L x .055 D x .055 P.
Iron-Clad

#6305 Micro-
Spade .500 L x
.055 D x .055 P.
Iron-Clad

#6312 Pencil
.750 L x .125 D x .010 P.
Copper-Alloy

#6313 Pencil
.750 L x .130 D x .020 P.
Iron-Clad

#6316 Pencil,
Offset .750 L x
.130 D x .020 P.
Iron-Clad

#6318 Long
Taper Chisel
.750 L x .125 D x .125 P.
Copper-Alloy

#6319 Long
Taper Chisel
.750 L x .130 D x .130 P.
Iron-Clad

#6324 Long
Taper Pyramid
.750 L x .125 D x .010 P.
Copper-Alloy

#6325 Long
Taper Pyramid
.750 L x .130 D x .020 P.
Iron-Clad

#6331 Screw-
driver .750 L x
.130 D x .130 P.
Iron-Clad

#6333 X-long
Taper Pencil
.750 L x .130 D x .035 P.
Iron-Clad

#6366 Long
Taper Chisel
.750 L x .211 D x .130 P.
Iron-Clad

#6368 Long
Taper Pyramid
.750 L x .187 D x .015 P.
Copper-Alloy

#6473 Long
Taper Chisel
.750 L x .375 D x .187 P.
Copper-Alloy

#6481 Taper
Screwdriver
.750 L x .130 D x .082 P.
Iron-Clad

#6482 Micro-
Spade Stepped
.500 L x .130 D
x .050 P.
Iron-Clad

Imperial Tips are designed to fit Imperial Heat Cartridges only.

(B)

Fig. 2-35. Soldering tip shapes and sizes. (B) Accessory tips. (*Courtesy* Ungar Div., Eldon Industries)

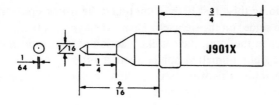

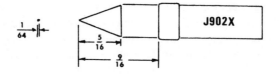

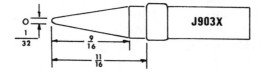

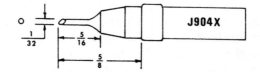

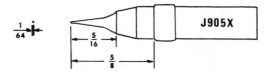

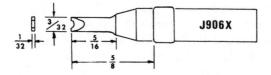

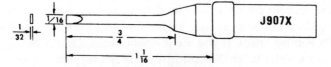

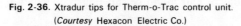

Fig. 2-36. Xtradur tips for Therm-o-Trac control unit. (*Courtesy* Hexacon Electric Co.)

try doing this: Just before you touch the iron's tip to the joint, melt some solder on the tip itself. Then, when you apply the wetted tip and solder to the joint, the capillary action of your premelted solder will make your unmelted solder melt a little sooner. You may waste some solder this way, but which is cheaper to replace—solder or a blown circuit or component?

If you have to solder in a small area where connections are close together, the secret is to modify the foregoing directions and solder *without using flux.* This may sound contrary to all that's been said before (except for the ultrasonic fluxless soldering), but there is sound reasoning behind it. To make a successful solder joint, the material to be soldered must be scrupulously clean. All traces of metallic oxide and dirt must be removed in order to expose the base metal. The resin flux core in electronic solder does just this, exposing the base metal and making it possible to wet the surfaces. The problem when soldering in small, confined spaces is that the flux in the molten solder wets *any* and *every* surface. The solder may then flow out and away from the connection you are making into adjacent areas. There is, however, one technique (used in the production of miniature instruments that contain circuit boards) that involves the use of *fluxless* solder. In this technique, the parts to be soldered are cleaned thoroughly and pretinned. When the solder is applied to the connection, it easily bonds the pretinned parts together. Owing to surface tension, the unfluxed solder forms a droplet that covers the connection, rather than flowing into the areas where it isn't wanted. You can use the same technique, both on new projects or for servicing in the field. First, pretin the parts to be soldered and the holes they are going into in the normal fluxed fashion. Then, melt a drop of solder on the tip of your low-wattage iron, as shown in Figure 2-37. After melting the solder, wait a few seconds for the flux to evaporate—for smoking to cease—then apply the tip of the iron to the connection, as shown in Fig. 2-38. The solder will flow off the tip and form a neat little ball around the connection and *nowhere else!*

After you have completed a soldering job on a printed circuit, really "eyeball" the foil carefully. Look especially for solder bridges or hairs between closely spaced foils. A magnifying glass or a magnifying lamp can be extremely helpful for this kind of checking (in troubleshooting, also). If you are in doubt about a solder bridge or hair, run a knife or an X-Acto blade through the area where you think the bridge or hair might be shorting. You won't hurt anything,

Fig. 2-37. Dissipating the flux.

Fig. 2-38. Letting the fluxless solder make the connection.

and you could be eliminating some future headaches. Finally, check all your connections with an ohmmeter, in a manner similar to that described earlier for hard-wired circuits.

3

Disconnecting/ Reconnecting Existing Circuits

Disconnecting and reconnecting existing circuits is just as much an art as making the original connections. Many older circuits are hard- and hand-wired, using globs of solder and stranded wire. Also, there may be many connections on a single terminal and, somehow, the one you want to remove is always on the bottom. Or some yo-yo has made a double-wrap mechanical connection, and that's the one you have to disconnect to test and/or repair. As for the globs of solder, they can melt and form little balls or splatters (Fig. 2-22) that fall onto some bare leads, making some nice little short circuits. And at the same time, while you're trying to get the stranded-wire leads off, when they finally do come loose they splatter hot solder all over your hands and the rest of the circuit.

With existing microcircuits, printed circuits, integrated circuits, and miniaturized components, disconnecting and removing parts present other problems. Many PC boards cost a fantastic amount of money to manufacture, yet you may have to remove an IC with some 14 to 20 leads (or more) without ruining the board, or the IC itself, because it may still be good or portions of it still usable. Thus, this separate chapter is devoted to disconnecting and reconnecting existing circuits.

Hard-Wired Circuits

If you've ever seen the underside of an old tube type of TV set, you know what kind of rats' nests hard-wired circuits can be. Wires seem to go everywhere and are often brittle, with their insulation cracked or peeling. Capacitors and resistors appear to hang in mid-air, and a 7- or 9-pin tube socket looks as though it has 20 leads going away from it, to say nothing of a tiny 0.25-watt resistor sitting inside the ring of socket terminals. Be of good heart—this is the worst you might encounter, and even this is conquerable with the proper tools and correct techniques.

Tools Required

To start with, you're going to need all the tools listed in Table 2 except, possibly, the wire-wrapping tool. On the other hand, some recent electronic devices use controls whose lugs have wire-wrapped connections. You can, of course, unwrap them with long-nose pliers, but you stand a greater chance of breaking off the lug and ruining the control.

In addition to the tools in Table 2, get some copper braiding, like that in 52-ohm antenna coax cable, and some solder paste. If that's not too handy, get a couple of rolls of "Soder-wick" or "Hex-Wik" at your nearest electronics parts store (with these you don't need the solder paste). They come in various sizes, for PC con-

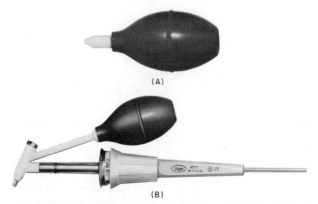

(A)

(B)

Fig. 3-1. Bulb-type solder "gobblers." (A) Model 7805 Solder-Off bulb; (B) Model 7800 Hot-Vac® desoldering tool. *(Courtesy* Ungar Div., Eldon Industries)

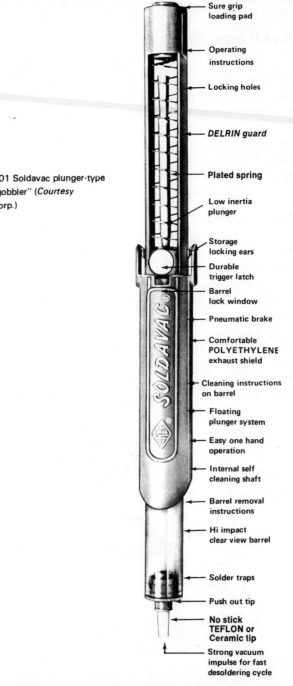

Fig. 3-2. Model 101 Soldavac plunger-type solder "gobbler" (*Courtesy* EREM Corp.)

Sure grip loading pad

Operating instructions

Locking holes

DELRIN guard

Plated spring

Low inertia plunger

Storage locking ears

Durable trigger latch

Barrel lock window

Pneumatic brake

Comfortable POLYETHYLENE exhaust shield

Cleaning instructions on barrel

Floating plunger system

Easy one hand operation

Internal self cleaning shaft

Barrel removal instructions

Hi impact clear view barrel

Solder traps

Push out tip

No stick TEFLON or Ceramic tip

Strong vacuum impulse for fast desoldering cycle

nections or multiple hard-wire connections. For the latter, there are also the solder "gobblers," shown in Figs. 3-1 and 3-2. Both use vacuum to suck up excess solder. Also, get some anti-wicking tweezers, such as those shown in Figs. 2-20A and 3-3. Last, if you haven't done so already, buy the soldering accessories shown in Fig 2-12 and the soldering aids shown in Fig. 2-19.

Desoldering Techniques

If you're working on older, hard-wired equipment, you'll notice that many of the joints are multiple-type with globs of solder on them. You'll also notice that, rather than one wrap of the wires (or leads) around the connection, each may be double or triple wrapped, making it difficult to see the precise wires (or leads) you want to disconnect. Here's where the "Soder-wick," "Hex-Wik," or the braid and paste combination comes in very handy, as do the anti-wicking tweezers. First clip the tweezers over any plastic-covered leads going to the junction, as shown in Fig. 3-3. Then place the "Soder-

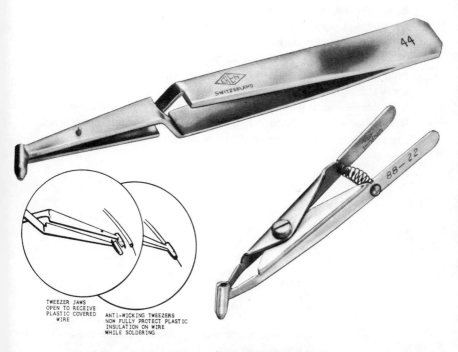

TWEEZER JAWS
OPEN TO RECEIVE
PLASTIC COVERED
WIRE   ANTI-WICKING TWEEZERS
              NOW FULLY PROTECT PLASTIC
              INSULATION ON WIRE
              WHILE SOLDERING

Fig. 3-3.   Anti-wicking tweezers. *(Courtesy* EREM Corp.)

wick"/"Hex-Wik" on the junction and apply the iron or gun to the "Soder-wick"/"Hex-Wik." (If you're using the coax braid, dip it into the paste before applying it to the junction.) In no time the excess solder will be sopped up as gravy is by a chunk of bread. Spotting the wire or lead you want in then a snap, and the anti-wicking tweezers have prevented the plastic insulation from melting. Also, once you've spotted your wire or lead, you can use the slotted end of your soldering aid to jiggle your wire or lead off the junction while you apply heat.

If you want to use the solder "gobbler," look at Fig. 3-4. This "gobbler" (shown close up in Fig. 3-1B) consists of a soldering tool, a collector tip, and a bulb. First, you squeeze the bulb, then place the collector tip on the junction. When the solder melts, you simply let go of the bulb. The suction pulls the liquid solder right into the bulb. You can force the solder out simply by squeezing the bulb again.

The procedure for using the "gobbler" illustrated in Fig. 3-2 is slightly different. Here you need both hands. First, you put the iron or gun on the junction and when the solder melts apply the "gobbler." Pressing down on the loading pad and then releasing it creates the suction to draw up the molten solder. To get rid of the solder, you press the loading pad again.

If the junction contains leads from solid-state components, such as a diode or transistor, use the heat-sinking tweezers to draw away the heat of the iron or gun, as shown in Fig. 2-21.

Sometimes it doesn't pay to desolder. Suppose you have a defective resistor or capacitor tied to a junction or a can (electrolytic capacitor) on one side and another permanent connection such as a

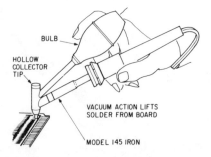

Fig. 3-4.   Using the Ungar No. 7825 solder "gobbler." *(Courtesy* Ungar Div., Eldon Industries)

tube socket pin on the other. For the latter, especially, it might be dangerous to try to desolder and remove the defective part. Trying to get the part off could easily break an old pin, and then where are you? To avoid this, simply cut out the offending component, leaving enough of the leads on both ends to allow you to connect the new part to them. If you make your connections as they're shown in Fig. 3-5, they'll be just as good as the real thing.

Speaking of tube sockets, desoldering and disconnecting must be done very carefully. Since they are tube sockets, the equipment is obviously pretty old. (We're not talking about the tube sockets on PC boards, which are a little more recent.) Thus, the socket terminals may have crystallized with age and become brittle. Also, manufacturers formerly had a tendency to double wrap any wires or leads connected to the terminals, making disconnection difficult. So your best bet when desoldering these terminals is to sop up all the excess solder you possibly can. Then take your subminiature cutters and grip the most convenient lead *end* on the terminal and gently pull it away while applying heat. (The cutters will give you a better grip than that of long-nose or duckbill pliers.) Once you've got the lead started, you can probably peel it off gently without applying the iron or gun—there's not enough solder to make it difficult. (This same procedure can also be used on the lugs of old controls; the same conditions prevail.) Once you have the offending lead off the socket terminal, gently ream out the hole in the terminal, using the iron or gun and an icepick, a piece of enamel wire, or even a needle held by your duckbill or long-nose pliers. The same type of reaming process can be used in the holes on electrolytic lugs, control lugs, and so on.

When desoldering from a direct chassis point, use a high-wattage gun or iron (Fig. 2-15)—the chassis acts like a heat sink, making your gun or iron less effective. Also, don't sop up any excess solder unless you have to; the older solder can be used for capillary action to make the new solder for your reconnection melt faster. Further-

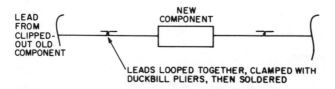

Fig. 3-5. Using old component leads to connect new component.

more, be sure to heat sink *any* and *all* component leads tied to that chassis point. The heat from the high-wattage gun or iron would otherwise cook the component—but good!

One final word about desoldering hard-wired equipment—be it old or new. When you're desoldering junctions that have *stranded* wire on them (and it's a common occurrence), be sure you've sopped up as much excess solder as possible. Then remove the wiring as gently as possible, with as little heat as possible. Stranded wire sops up solder, and if you whip it around during desoldering, it could spray solder all over the place, possibly into your eyes and definitely on your hands. Also, that spray could land between two socket terminals or some other close junctions and form a bridge, causing a short you might have to spend hours looking for. So it is *better* to be *careful!*

Printed Circuits

Printed circuits present problems different from those of hard-wired circuits. PC boards normally have only one lead or wire per hole, so it's fairly easy to remove excess solder. On the other hand, many components (e.g., a transformer) are bolted down on a hard-wired chassis, making them simple to remove after the leads have been desoldered. The same transformer on a PC board has to have its leads desoldered simultaneously to enable its removal. But don't be too alarmed; the problems are not that complex.

Tools Required

The tools listed previously as being necessary for hard-wired desoldering are also necessary for micro and standard PC desoldering, except that no high-wattage iron or gun is required. The same types of irons required for soldering micro and standard PC boards (i.e., 4.5-6 watts and 10-20 watts) are all that are necessary. However, the irons should be capable of accepting the screw-in type of desoldering tips shown in Fig. 3-6. Tips such as these are a must if you want to do the job neatly, efficiently, and with a minimum amount of breakage. The IC insertion and extraction tools illustrated in Figs. 3-7 and 3-8 are also helpful.

Desoldering

The first thing to remember is to reread the portion of Chapter 2 that refers to soldering of microelectronic equipment. The rules that hold there for soldering hold also for desoldering—maybe more

Fig. 3-6.   Desoldering tips.

Fig. 3-7.   Princess Model No. 6982 DIP IC extraction tool. *(Courtesy* Ungar Div., Eldon Industries)

Fig. 3-8.   Princess Model No. 6983 TO-5 IC extraction tool. *(Courtesy* Ungar Div., Eldon Industries)

so. Somehow, in desoldering there is a tendency to hold the hot iron or gun on the joint just a little longer than in soldering. If you reread or remember the cautions of temperature, tip size, tip dwell, etc., discussed in Chapter 2, your desoldering will be that much safer and easier.

The next thing to remember when desoldering micro and standard PC boards is to use the "Soder-wick"/"Hex-Wik," or braid and paste combination (mentioned previously for hard-wire desoldering) to sop up excess solder on any joint before attempting any parts removal (see Fig. 3-9). Do this if you can get at the foil side of the board; sometimes you can't. Don't fret; that problem is soluble, too.

Let's start with the simple stuff, such as removing a two-lead part—for example, a resistor or diode. First, check the foil side of the board to see if the leads are bent over. If they are, either bend them up straight or clip them off, using your subminiature *cutters* in either case. If you have to, use your iron to melt any residual solder while bending or clipping. (If you're removing only *one* lead, say, to check the out-of-circuit resistance or to see if a two-lead capacitor is shorted or open, *bend only,* don't clip. You might not be able to solder back the part if it's still good.) Next, grab the most convenient

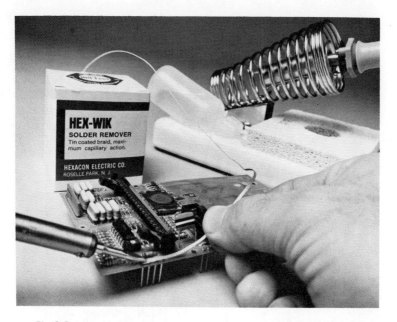

Fig. 3-9. Desoldering using commercial braid. (*Courtesy* Hexacon Electric Co.)

part lead with the slotted end of your soldering aid (Fig. 2-19), long-nose or duckbill pliers, and gently pry or pull it out as you apply heat to the foil side. (If you can't use either of these tools, simply insert the other end of your soldering aid or the tip of a small, flat-bladed screwdriver under the part and pry it out as you apply heat.) When the lead comes out, *slide* your iron away from the foil. Pulling it away perpendicularly may cause the foil to peel off because the heat has broken the bond between foil and board. Now repeat the procedure for the other lead.

If by some chance you've broken the foil during parts removal, don't despair. Just be sure the remaining foil is clean. Then when you insert the new part, simply bend over the corresponding lead so that it touches the foil and solder. Cut off any excess lead. On the other hand, if the foil was already broken and no new part is going in, you can still fix the break. Just lay a piece of solid hookup wire over the break along the foil so that each end of the wire extends at least ¼ inch beyond the break. Now solder, and the connection will be as good as new (see Fig. 3-10).

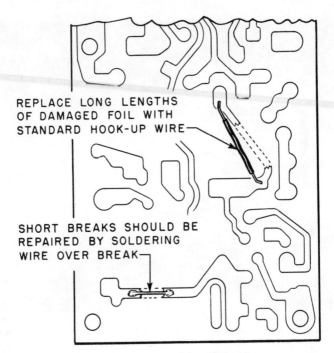

REPLACE LONG LENGTHS
OF DAMAGED FOIL WITH
STANDARD HOOK-UP WIRE

SHORT BREAKS SHOULD BE
REPAIRED BY SOLDERING
WIRE OVER BREAK

Fig. 3-10. Repairing the foil on PC boards.

Now comes the problem of desoldering or soldering where you can't get at the foil side of the board. If the part to be removed and replaced is flush to the board, and many are, you have only to remove the board and desolder as already described. If, however, the part is elevated off the board, as in Fig. 2-31, the task is easier. If you can cut the leads and still have enough left to bend them as shown in Fig. 3-5, do so. If not, take your duckbill or gas pliers and pulverize the part until only the bare leads remain. Then connect the new part as in Fig. 3-5 and solder, using a heat sink, to make sure the old solder on the foil side doesn't melt also. One warning: If you're replacing a diode, be sure to notice the polarity before you remove or pulverize the old one; the same holds true for polarized capacitors.

Sometimes a part is easy to take out but tough to replace because of the equipment configuration. So if you have a little room on the foil side, connect the part on that side by placing the leads along the appropriate foil and soldering them on. Use a piece of masking tape to hold the part in place while you're soldering the first lead; then remove the tape.

Now that you've removed the part, is the hole visible or clogged with solder? If the latter condition exists, take a wooden toothpick and place its end where the hole is supposed to be. Then heat the foil again while pushing down *gently* on the toothpick. Once the toothpick goes through, twirl it around as you remove the iron, then remove it (again *gently*) to prevent pulling the foil away from the board.

One of the problems involved in desoldering printed circuits is locating the right foil path to melt. As soon as you look on the foil side, one conductor looks like another. You can solve your problem with a bright lamp, a flashlight, or even a penlight held close to the part side of the printed circuit. The light will make the board translucent, and you can easily locate the right foil connection.

What do you do when you have to desolder a part with eight connections from a printed circuit? One method is to remove all the solder from each connection individually until all the leads are free. There are various ways to practice this technique. One requires holding the foil above the soldering iron so that gravity will help the melted solder flow down from the connection to the iron. A second way is to suck the excess melted solder from each connection with a "gobbler". Another alternative would be to use the "Soder-wick"/"Hex-Wik" or braided shield wire for a solder sponge.

Another method that can be successful, but that generally puts a lot of strain on the part and the board, is the heat-and-lift technique. One or two connections are heated while a lifting force is applied to that section of the part. The heated connections will generally pull up slightly. Heat two or three more connections while applying force to that section of the part. Keep on doing this around the perimeter of the part and eventually the leads will lift out of the printed circuit. This technique works best on components with four leads or less spaced rather far apart.

A fourth method used to remove parts with multiple connections is heating all the soldered joints simultaneously and lifting out the part. Here's where those desoldering tips illustrated in Fig. 3-6 may be worth their weight in gold. Just use the one most suitable for your application (but not so large that it might contact a heat-sensitive part that does *not* have to be removed), and your problem is eased considerably. When you use these tips, make sure that the part leads have not been bent over or crimped before soldering. If they have been, you will have to straighten each pin out individually and snip it off before the desoldering tips can be used successfully. Also, make certain that the tip has reached its operating temperature, and tin the tip heavily so that maximum surface contact between the tip and all the part leads can be made. If you make good initial contact with all the leads, it will take only a second or so before the solder has melted and you can lift the part from the board. If the part you're extracting happens to be an in-line (DIP) integrated circuit, you can use the extractor shown in Fig. 3-7 to extract the IC while you're desoldering with a long, rectangular tip such as the one shown in Fig. 3-6. If, on the other hand, you're extracting a TO-5 type of IC, you can use the extractor shown in Fig. 3-8. Now you can pull out the IC while desoldering without burning your fingers. As for replacing the DIP integrated circuit, you can use the insertion tools shown in Fig. 3-11. If you're careful, you can use a socket wrench (or plastic nut starter) to insert TO-5 types. Just get the size socket that fits snugly over the IC, then use the wrench to guide the IC into the correct holes. Be very careful; jamming the metal socket over the IC could keep the IC from coming loose, or worse, you could crush the IC and its internal parts.

One warning about using the desoldering tips shown in Fig. 3-6. If you're working on a small, closely packed wired circuit like that of a microprocessor, I advise against the use of these tips. The heat they generate could both desolder other circuit portions without your realizing it and destroy some nearby circuits and/or components. Better to use the "Soder-wick"/"Hex-Wik," or braid and paste method, as shown in Fig. 3-9.

All these desoldering methods work more or less. Try to figure out which one is the best for a particular application.

If, during the course of your desoldering, you should run into some wire-wrapped connections (probably on control lugs) that must be removed, use great care in removal. First, you might break off the wire or lead and find it to be too short for reconnection. Second, you might break off the terminal around which the wrap is made. So, to be on the safe side, do your unwrapping with an unwrapping tool, such as that shown in Fig. 2-4A. If that's not available, start the unwrap with your subminiature cutters, then continue *slowly* with your long-nose or duckbill pliers.

PART NO.
A23-2050

PART NO.
A23-2050P

Fig. 3-11. DIP insertion tools with 14 and 16 pins. Part No. A23-2050P is
a gold-plated version for use with MOS ICs where the leads
must be shorted during insertion. *(Courtesy* JERMYN)

Soldering and Reconnecting

Just about everything concerning soldering and reconnecting in
existing circuits has been covered in the previous material on desoldering and in soldering in new circuits. However, there are still a
couple of tips that apply.

A major point in soldering old circuits is to make sure the joint
is clean *before* you apply the iron and new solder. Don't depend on
the resin flux to clean the joint. Use the wire brush end of your soldering aid (Fig. 2-19) instead.

While desoldering and/or disconnecting one part, you may in-
advertently cut off a piece of enamel wire from some transformer. If
so, be sure to remove enough enamel from the remaining wire to assure its taking solder. One way to do this is to heat the wire end with
the flame of an ordinary book match. Hold about ½ inch of the wire
in the flame, and as soon as you see the flame flare, remove it. Then
burnish the ½ inch of wire with the component-lead cleaner shown
in Fig. 2-12, or with fine sandpaper, and your ability to make the
connection is assured.

If you have to reconnect to a tube socket that has one or more leads already attached to it, leave the socket terminal alone. Simply twist the lead you have to reconnect around one or the other of the existing leads, and solder. Just be sure the existing lead is not enameled.

Even though you've been cautioned earlier about leaving parts hanging in midair after soldering, in older equipment this is often not feasible. In fact, it's highly possible that an original part you are replacing was hung in midair during manufacture. You can do either of two things to remedy the fault. First, and best, try to find two firm anchor points with the required continuity that are sufficiently close together to permit anchoring the new part there. Or second, before you splice in the new part, put sleeves on both leads of the part; you can use ⅛-inch plastic tubing. Then either clamp the new part in as shown in Fig. 3-5, or, better yet, twist the ends together in a standard splice (see Fig. 3-12). Make the splice as short as possible and the sleeve as long as possible. Solder the splice, clip off the two ends sticking up (or down), then slide the sleeve over the still-warm splice. The heat will tend to mold the sleeve around the splice, holding it in place and reducing the chances of midair short circuits.

If you've unwrapped a wrapped connection, don't try to rewrap the connection—bending and rebending are not conducive to the health of metal. Simply follow good soldering practice (i.e., a single bend) and solder the wire or lead back where it belongs.

You may have to make solder connections that are almost impossible. They are buried under wires or parts deep inside a chassis. Try to move the blocking wires and parts away from the area as much as you can without putting excessive strain on their leads. A soldering gun with a built-in light is really helpful for this kind of soldering. If you don't have enough room to heat the joint and apply solder at the same time, first tin the lug and then tin the part's lead. Wrap the lead around the lug and then heat the joint until the tinned solder melts on both parts and blends together.

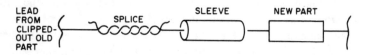

Fig. 3-12. Making a standard splice connection.

4

Troubleshooting

This chapter deals with generalized troubleshooting and repair procedures on the level of the advanced hobbyist and professional technician. In other words, it is written for those who have a good knowledge of tools, who can interpret a block diagram, and who can read a schematic diagram.

The following information primarily emphasizes PC solid-state troubleshooting. This is not meant to consign vacuum tube troubleshooting to the grave. The fact is, 50 percent or more of vacuum tube troubles are soluble merely by pulling out the old tube and putting in a new one (except in series-string circuits, unless you just happen to have the exact tube complement you need). If that doesn't do it, then you have to begin a real troubleshooting procedure to isolate the fault. The basic procedure is the same for both vacuum tube and solid-state circuits, with three major exceptions: (1) the components in vacuum tube circuits are larger, (2) the voltages and currents in vacuum tube circuits are larger, and (3) you don't have to desolder a tube to remove it. For example, a grid voltage may run anywhere from -7 V to, say, -10 V or even -15 V. On the other hand, the bias voltage between the base and emitter may range only from 0.05 V to 0.7 V, calling for a sensitive (low-scale) vacuum tube voltmeter (VTVM). As for part size, you may need an iron of higher wattage for the tube circuits.

Keeping these facts in mind, you will see that the following information applies equally to tube and solid-state circuits, unless otherwise stated.

Basic Troubleshooting Techniques

Prerequisites

Before going any further, one thing must be understood: A logical approach is required in order to isolate and correct any fault. For example, it would be foolish to try to repair a TV set without first turning it on to see what the symptoms are. How do you approach a troubleshooting job logically? The first thing to examine is whether you have the necessary technical know-how. Ask yourself the following questions.

Do you know how the equipment operates normally? This knowledge can come either from experience or from instruction manuals and manufacturers' data, or a combination of both. Also, no matter what experience you have, it cannot possibly encompass every existing circuit, and studies of data sheets for expected waveshapes, signal levels, resistances, voltages, and so on, must be made before attempting any trouble location and repair. Last, if you don't have much experience and the data available are sketchy or nonexistent, the best troubleshooting you can do is to realize your shortcomings and refer the job to some agency that can handle it.

Do you know what all the controls and adjustments are supposed to do and how to operate and adjust them? This, too, is based on experience, especially on a piece of simple equipment such as a radio or a black-and-white TV set. Even for these you simply do not adjust intermediate frequency (IF) stages unless you have data sheets telling you how much and in what sequence. As for complex equipment, such as a good stereo amplifier or a color TV receiver, control and adjustment settings are critical, especially if the components have been in use for a while.

Do you have available and do you know how to use the proper test equipment? A simple 1,000-ohm/volt voltohmmeter (VOM) may be fine for checking vacuum tube circuits. For checking transistor biases, which may range anywhere from 0.05 V-0.7 V, you really need a transistor checker, such as that shown in Fig. 4-1. Why a special tester? Well, not only does that simple VOM load down the circuit so as to give false indications, but its internal batteries may provide enough voltage to burn out the very component being checked. Thus, for troubleshooting a solid-state component, and especially integrated circuits, you need more sophisticated test equipment. Basi-

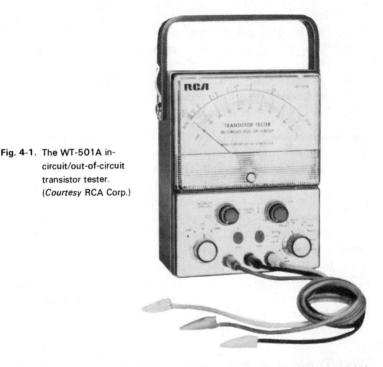

Fig. 4-1. The WT-501A in-circuit/out-of-circuit transistor tester. (*Courtesy* RCA Corp.)

cally, you need a good VOM, such as a Simpson Model 314 solid-state VOM and/or a Triplett Model 801 solid-state VOM. The major feature of both instruments, as far as we are concerned, is that they dissipate only microwatts (70 μW for the Simpson, 140 μW for the Triplett), making them safe for both in-circuit and out-of-circuit IC tests. In addition, both can measure dc voltages as low as 50 mV (millivolts); for ac ranges the Triplett can go as low as 5 mV, the Simpson to 10 mV.

You could, of course, opt for the digital voltmeter (DVM). A good DVM offers the following features:

1. Numerical readout
2. Pushbutton selection of functions and ranges
3. Automatic overload and protection
4. Automatic polarity indication
5. Automatic switching decimal point
6. High input impedance

7. Portable or ac powered
8. Internal calibration checks
9. Both ac and dc voltage ranges—millivolts to hundreds of volts
10. Both ac and dc current ranges—microamperes to amperes
11. Reistance ranges from ohms to megohms
12. Broad frequency response

Digital voltmeters are a little expensive for the ordinary pocketbook, but the features involved may well be worth it. And if you're really serious about troubleshooting, you'll also need a variety of signal sources—audio generator, RF generator, marker generator, logic pulser and so on, plus a good oscilloscope.

Do you have the proper tools and know how to use them? Most repairs can be made with the basic tools, such as those given in the previous chapters.

Once you have noted the symptoms of the trouble, are you able to analyze them before you plunge ahead? You must be able to think and have the patience to sit back and reason out where the trouble most likely lies. In simple terms, don't rush in to replace the speaker just because there's no sound.

Basic Troubleshooting Sequence

There are four steps in the basic troubleshooting sequence:

First, analyze the symptoms. Determine whether you understand all the symptoms, that is, whether they indicate normal/abnormal/subnormal operation or complete failure. For example, is the sound from one channel of a stereo receiver louder than from the other, or is one channel distorted, or is one channel (or both) completely without sound? Then, unless you have been servicing stereo equipment for a while, check all your available service data. They might contain the very remedy you need. (Also check the most obvious: Is the equipment plugged in and properly hooked up?)

Second, localize the trouble to a specific function or module. For example, if the symptom is "no picture" in a TV set but there are a raster and sound, you can be fairly sure that the trouble is not in the high-voltage, power supply, sweep, or audio sections, or in the picture tube itself. In other words, you have localized the trouble to either the IF or video sections. Keep in mind, however, that localization is not always definite: Failures in one section can often show up as abnormal symptoms in another section. As an example, a "fuzzy-sounding" speaker is often not the fault of the speaker itself but of a bad audio-output stage or weak input signal.

Third, localize the specific malfunctioning circuit within the localized function or module. This can be done with the power on by signal tracing or making voltage checks. Or it can be done with the power off by signal injection and/or resistance tests. Signal tracing, (Fig. 4-2) is performed by examining the inherent signal at a test point with a VOM, oscilloscope, or even a loudspeaker. In tracing, the input probe of the indicating device is moved from point to point, while the signal source remains a constant. On the other hand, signal injection (Fig. 4-3) involves injecting an artificial signal from some generating device(s). Here, the indicating device (the SPKR in Fig. 4-3) remains fixed at one point, while the signal-injection device is moved from point to point until the faulty circuit is found.

Such testing must be performed with the aid of a schematic diagram that shows waveshapes and/or voltage levels at the points being checked. Otherwise, test results can be meaningless. The

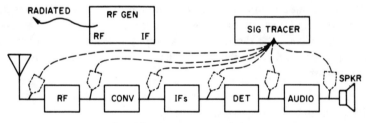

Fig. 4-2. Signal tracing technique.

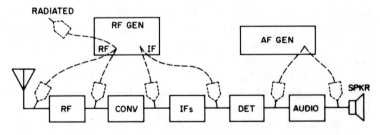

Fig. 4-3. Signal injection technique.

trouble may lie in a signal path. Power supplies neither contain nor propagate signals, hence signal injection and/or tracing are a waste of time. Here, only voltage and resistance measurements are feasible.

Fourth, localize the faulty component. Once you have localized the defective circuit, the first thing to do is to make a visual check of all the components involved. Check for loose connections, cold solder joints, unsoldered joints, cracked resistors, resistors with strange color bands (which indicate overheating), bulging paper capacitors, cracks in PC wiring, and the like. You should also look for burned resistors, although in solid-state circuits, with their low voltages (except for power supplies and audio output stages), such occurrences are rare.

Next, check voltages against those called for in the schematic diagram. (If the original symptoms were burning or arcing, this procedure is not feasible. In such cases, the burning or arcing will do your narrowing down for you). Voltage checks are generally more effective than are resistance checks of solid-state equipment. For example, checking a capacitor that is in parallel with a low-resistance resistor may lead you to think you have a shorted capacitor. Or if you are using an ohmmeter to check an emitter resistor that has its other end tied (directly or indirectly) to the base, what you might be measuring could be the emitter-base forward resistance since, with the applied voltage of the meter, the transistor could become forward-biased.

Miscellaneous Suggestions

If you are making in-circuit repairs, pull the plug. Many of today's solid-state circuits are the "instant-on" type, meaning that some circuits are "live" even though the power switch has been turned off. And, aside from getting a nice shock, the transient surges in electric current you may get while replacing a part could ruin the part.

If the equipment operates from an ac line, check the line first. Low line voltage can cause all sorts of trouble, especially in critical IF and phase-inversion circuits.

Resistance measurements are usually not necessary in solid-state equipment, except to check for open circuits in transformers and coils. (Because of the low-voltage power supplies used, resistors have little tendency to burn up or change value.) Since the ohmmeter contains a battery, wrong polarity voltage may be applied to a circuit stage and damage the transistor permanently. So remove the transistor or component from the circuit before attempting resistance measurements.

Alignment is an excellent technique when checking RF and IF stages for sensitivity. A shorted turn (solder splash?) in an RF or IF coil is quickly identified by poor response when tuning the coil. For example, rotating the slug in an IF coil one-half turn normally has a great effect on the sensitivity of a transistor tuner. If this adjustment produces little or no effect, the coil is either defective or the associated stage is inoperative.

When working with powered-up solid-state equipment, make sure all circuit parts, such as speakers or yokes, are connected. If the load is removed from some solid-state circuits, heavy current will flow, resulting in damaged integrated circuits.

If a transistor element appears to have a short, check the settings of any operating controls associated with the particular circuit. A volume control set to zero, for example, can give the same reading as a short from base to ground.

Guard against shorts. A short between the collector and base of a transistor may damage it and, often, other transistors associated with it, as in the case of direct-coupled stages or those in a power amplifier output stage. Such shorts occur even in the time it takes for a dropped screwdriver to glance off a pair of socket terminals or to short a lead or terminal to the chassis.

A transistor may be damaged if its base is placed at or near the collector potential. Therefore, make sure the base leg of the biasing circuit is not open on the emitter side.

When replacing power transistors, be sure there are no metal shavings on heat sinks or mica insulators that might cause shorts or prevent adequate heat dissipation. Use silicone grease between heat sinks and transistors, as well as on both sides of mica insulators, for better heat conduction.

Be sure there are no leakage paths to the ac line through test equipment or soldering irons, since line voltage applied across two transistor terminals could cause breakdown.

When you are checking capacitors in-circuit, do not do so by jumping a new one across the old one. The transient surge that occurs when you touch the new one into the circuit might be just enough to damage or ruin an integrated circuit (or semiconductor) in that circuit.

Here are some additional hints on capacitors in IC circuits. If you have a failed IC, check to see whether the circuit is using capacitors for input coupling to the IC, for output coupling from the IC, for noninverting feeback, or for output-line loading. Capacitors, especially if they are low-impedance types, have a habit of presenting instantaneous current levels at the input/output of a device that are greater than the rated dc levels. Although a good IC should handle these peaks for a while, the peaks cause premature aging and early breakdown. The really bad feature is that the inclusion of capacitors in this manner is a design error or cost-cutting procedure, so all you can do is replace the integrated circuits, or figure out a better design for that circuit.

Check to see whether a failed IC is driving some sort of coil or incandescent lamp. Here, again, the designer may have decided to cut corners by using the IC's maximum rated output current as his driving source. This also ages the IC prematurely, again necessitating frequent replacement or your own redesign.

Last, when there is no obvious cause for a low voltage at some point in the circuit or an abnormally high resistance exists, take a magnifying glass and look for cold solder joints, PC wiring breaks or hairline cracks, or even tiny bits of solder (solder bridges) that might be shorting a couple of closely run PC wires. Breaks or cracks can be repaired by the judicious use of a little solder or even by a discrete piece of wire, if the space is available (see Fig. 3-10).

Cold solder joints can sometimes be found by using an ohmmeter. Remove all power. Connect the ohmmeter across two wires leading out of the suspected joint, as shown in Fig. 4-4. Flex the wires with the ohmmeter probe tips. Switch the ohmmeter to different ranges and check if there is any change in resistance. For example, a cold solder joint can appear to be good on the high ohmmeter ranges but appear as an open joint on the lower ranges. Look for resistance indications that tend to drift or change when the ohmmeter is returned to a particular scale.

A little thing that sometimes makes PC test measurements difficult is the type of connector or prod in test leads. Generally, there is not enough surface area on the parts' leads to allow the use of an alligator clip. If you do manage to clip it on, the slightest movement

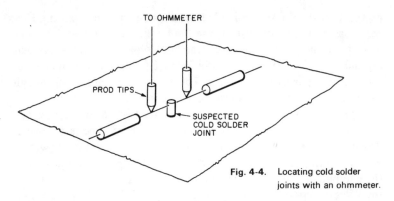

Fig. 4-4. Locating cold solder joints with an ohmmeter.

will cause the clip to release and maybe fall into the circuit, causing additional troubles. You can use leads with minigator clips, but even these small clips are too large to be really effective in most printed circuits. The best solution is to use test leads terminated with special miniature insulated hooks, as shown in Fig. 4-5. These hooks are made especially for PC testing and work like a charm. Each con-

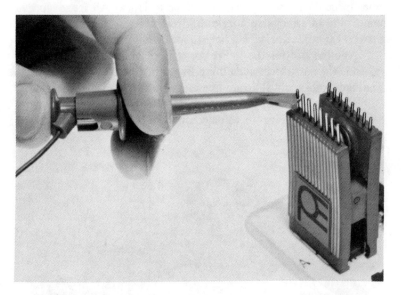

Fig. 4-5. Using the miniature insulated hook with a Model 3916 "Dip-Clip" IC test clip. *(Courtesy* Pomona Electronics)

nector consists of a spring-loaded hook inside a plastic housing. Finger pressure is required to expose the hook. After the hook is connected and the pressure released, the hook will retract, locking itself around the wire or testpoint and being completely insulated from other nearby parts or conductors. These connectors are small enough to reach and hook the smallest PC component lead.

Got intermittent problems? There are two general types: one caused by heat, the other caused by lack of heat. The latter is evidenced by intermittent malfunction until the equipment really warms up. Most intermittents are caused by heat. The problem arises when the equipment gets fully warmed up. A fairly good solution to this is the "freeze" spray, obtainable in aerosol cans. To use such a spray, let your equipment warm up until the intermittent occurs. Then spray each component in the suspected circuit individually. Eventually, you'll hit the right one and the problem will disappear. Double check, though, by letting the component warm up again to see if the problem recurs. Have patience; it could take a while. Then spray again, and if the problem goes away again, replace the bad part. While you're spraying, make sure the spray doesn't hit a hot tube—it could cause the glass envelope to crack. All freeze sprays come with an attachable nozzle that resembles a long piece of spaghetti. Using this nozzle will help to localize the spray.

Another use for the freeze spray is checking for cracks in the PC foil. If you suspect something like this, just spray the doubtful area; any cracks will show up like magic in the resulting frosting.

If you have the opposite problem—lack of heat—get a heat gun like that shown in Fig. 4-6. (You could use a hair dryer, but its nozzle

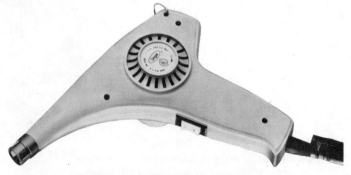

Fig. 4-6.   Model 6966C three-wire heat gun.
(*Courtesy* Ungar Div., Eldon Industries)

is too wide for accuracy.) Turn your equipment on, wait for the problem to appear, then direct the gun's heat at the suspected circuit. When the problem goes away, you've got the spot localized for part(s) substitution. The gun can also be used to determine whatever you have wrong: cold solder joints, PC cracks, loose connections, and so on. A little judicious heat will make the suspected area(s) expand, curing the ill temporarily. If none of these is the culprit, you haven't wasted your time: You're now pretty sure your problem is elsewhere. By the way, the heat gun illustrated comes with various tips for directing the heated air more accurately and/or for heat shrinking plastic tubing.

Last, it just may be that you're working on older equipment without benefit of a schematic diagram. The possibility exists, therefore, that you may find a bad resistor or capacitor, and you can't determine its value because you've forgotten your color coding. Figure 4-7 has been included here just to cover that contingency.

Solid-State Troubleshooting

Aside from integrated circuits, which contain a multiplicity of circuits, the two major semiconductor devices are diodes and transistors. (In monolithic integrated circuits, diodes are often formed by laying down a transistor pattern wherein the collector and base are shorted together to form the cathode of the diode, with the emitter forming the anode.) Since it is simpler to test diodes, we will begin with them. However, whether you are testing a discrete or an IC diode, the procedures are the same.

Solid-State Diodes

Diodes have three basic requirements: (1) They must be able to pass current in one direction (forward current) while preventing or limiting flow in the opposite direction (reverse current); (2) for a given forward current the voltage drop across the diode should not exceed a specific value; and (3) for a specific reverse voltage, the reverse current should not exceed a given value.

All the foregoing are best determined out-of-circuit, using an oscilloscope and appropriate data sheets or a diode tester. However, an ohmmeter can be used to good effect in checking a diode's current-passing or current-limiting capabilities. A good diode will exhibit high resistance in the reverse direction and low resistance in the forward direction.

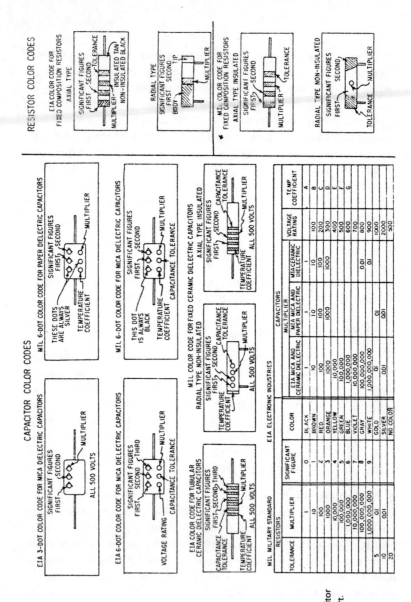

Fig. 4-7.
Resistor/capacitor color-code chart.

If the resistance in the reverse direction is low, the diode is probably leaking. If it is low in both directions, the diode is most likely shorted. If high resistance is found in both directions, the diode is probably open. The basic factor in determining a good diode, in addition to high-reverse/low-forward resistance, is the ratio between the two resistances. The actual ratio depends on the type of diode. For small signal diodes the ratio should be at least 1:100, forward-to-reverse (front-to-back ratio is the more common terminology). For power diodes, a satisfactory ratio is at least 1:10. For IC diodes, the ratio should be at least 1:30. Incidentally, to check IC diodes you must have a schematic diagram of the IC itself, since most equipment schematic diagrams show an IC simply as a sideways triangle.

You can also determine whether a discrete diode is a germanium or a silicon type when checking the forward resistance. A germanium diode will start conduction when about 0.25 V is placed across it. A silicon diode needs more voltage, about 0.6 V, to cause it to conduct. Just measure the voltage drop across the diode with a voltmeter while checking the forward resistance with an ohmmeter, as shown in Fig. 4-8.

A VOM can check zener diodes in the forward and backward direction, but it cannot check the zener breakdown point. Zeners will check exactly the same as a conventional diode: low resistance in the forward direction and high resistance in the reverse direction. If the zener doesn't check correctly in this manner, you know it's bad. However, even if it does check correctly, you still don't know if it will break down at the correct zener point.

Figure 4-9 shows a simple but effective circuit for checking zener diodes. The zener is reverse biased to its zener point by the current flowing through the 2 k pot. At the zener's specific current level, the diode will break down, causing the voltmeter voltage to remain constant throughout the wattage range of the zener. For example, suppose the zener is rated for 10 V, 400 mW. When the zener point is reached in the test circuit, the diode will conduct for any voltage above 10 V, keeping a constant 10 V across itself and being measured by the voltmeter. The voltage will remain 10 V as long as the current remains below 40 mA (400 mW). As a rule of thumb, remember that zeners will not operate properly unless loaded to 20 percent of maximum load. In this case good zener action will begin when the zener current reaches around 8 mA.

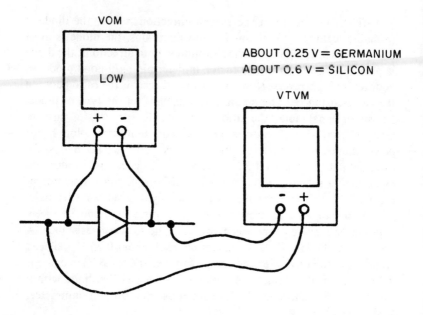

ABOUT 0.25 V = GERMANIUM
ABOUT 0.6 V = SILICON

Fig. 4-8. How to determine if a diode or rectifier is germanium or silicon.

Did you ever have to replace a defective zener and find that you couldn't locate the exact replacement? Zeners can be connected in series to increase the breakdown point so long as you do not exceed the current rating and power dissipation of either zener. If you need an 18-V zener and have available only an 8-V and a 10-V zener, sim-

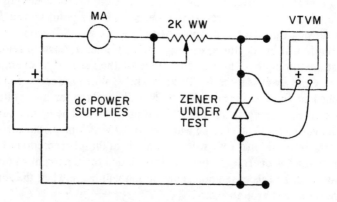

Fig. 4-9. Test setup to check a zener diode.

ply connect them in series. If their wattage rating is the same or higher than the original, they will work just as well. Three 6-V zeners in series would be the electrical equivalent of an 18-V zener.

Another way to change a zener voltage is to connect germanium or silicon diodes in series with the zener. A germanium diode will increase the zener breakdown voltage about 0.25 V and a silicon diode about 0.6 V.

Transistors

The two most common types of transistors, be they discrete or integrated, are the bipolar (base, emitter, and collector) and the unipolar or field-effect (gate, source, and drain) types. We will begin with the bipolars.

Bipolar Transistors. Basically, linear circuits consist of resistors, capacitors, and so on, and bipolar transistors, as previously stated. The following discussion covers the way to track down troubles to these transistors and the way then to check the transistors themselves. Again, remember that on a schematic diagram, the symbol for an integrated circuit is usually a sideways triangle, as shown in Fig. 4-10. The dotted lines have been inserted by the author and

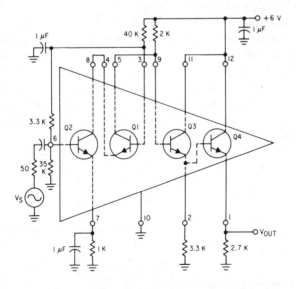

Fig. 4-10. Schematic diagram for a CA3018 Cascode Video Amplifier (dotted lines inserted for use in this text only). *(Courtesy* RCA Corp.)

are not normally given in the schematic diagram. So as with IC diodes you must have the IC schematic diagram if you want to check out an IC transistor.

Bipolar transistor circuits are best tested by voltage checks, using the applicable schematic diagram and a sensitive VOM, such as a Simpson Model 314 or a Triplett Model 801.

Figure 4-11 shows the basic connections for both pnp and npn transistor circuits; the coupling and bypass capacitors have been omitted for simplicity. In practically all transistor circuits, the emitter-base junction must be forward biased to obtain current flow through a transistor. In a pnp, this means that the base must be made more negative than the emitter. The emitter-base junction will then draw current, causing heavy electron flow from collector to emitter. In an npn, the base must be made more positive than the emitter for emitter-to-collector current flow.

The most common way to measure transistor voltages is between ground and the element; manufacturers' data generally specify transistor voltages this way. For example, all the voltages for the pnp of Fig. 4-11 are negative with respect to ground. The following discussion demonstrates how voltages measured at the transistor elements can be used to analyze failure.

Assume that the pnp circuit of Fig. 4-11 is measured and the voltages found are those of Fig. 4-12. The first clue that something is wrong is that the collector voltage is almost the same as the source voltage at R3, indicating that very little current is flowing through R3. The resistor could be defective, but the trouble is more likely caused by a bias problem. The emitter voltage depends mostly on the current flowing through R4; thus, unless the value of R4 has changed drastically, the problem is one of base bias.

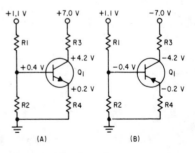

Fig. 4-11. (A) Npn and (B) pnp transistor circuits showing normal voltages.

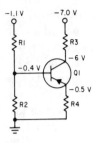

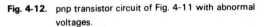

Fig. 4-12. pnp transistor circuit of Fig. 4-11 with abnormal voltages.

The next step, then, is to measure the voltage at R1. If the bias source voltage is, say, −0.9 V instead of the required −1.1 V, the problem is obvious: There is probably a defect in the power supply. If the bias source voltage is correct, then either R1, R2, or Q1 is defective.

The next step is to power-down the equipment and measure R1 and R2. If either is incorrect, there's your solution. If both values are correct, check R4 just to be sure. However, it is more likely that Q1 is faulty. (A short note on the foregoing: The capacitors are not shown in Figs. 4-11 and 4-12. So, before you pull Q1 or the integrated circuit it is part of, check out any capacitors that might be across R1-R4).

Having reasonably assumed that Q1 is faulty, there is a way to make sure: Fig. 4-13 illustrates the test setup if you do not have a transistor tester. With the voltmeter connected and the circuit operating, measure and notice the difference potential, emitter-to-collector. Now shut off the power and short-circuit the emitter-base junction, as shown by the dotted-line jumper. Reapply power and remeasure the emitter-collector potential. If it is not much higher than the original measurement, the transistor is defective.

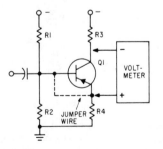

Fig. 4-13. In-circuit test setup.

Transistors (discrete) can also be checked out-of-circuit by simple resistance measurements (again assuming you do not have a transistor checker). Bipolar transistors are either pnp or npn types. This means that there are two pn junctions in each type. And since a pn junction is, in effect, a semiconductor diode, it can be tested in the same way. So to check an npn type, the ohmmeter leads are first connected as shown in Fig. 4-14A. Then the leads are reversed so that the meter COM (common) lead is connected to the base and the positive (+) lead is connected to the emitter. The ratio of front-to-back resistance should be at least 1:30. If this step checks out, the process is then repeated as demonstrated in Figs. 4-14B and 4-14C, each time reversing the leads to determine the front-to-back ratios.

As for pnp types, the procedure is exactly the reverse. That is, if it were a pnp in Fig. 4-14A, the meter COM lead would be connected to the base and the + lead to the emitter, and then reversed.

Unipolar (Field-Effect) Transistors. Unipolar (field-effect) transistors (FETs) can be checked in a variety of ways. A good VOM will tell you whether the proper source and output voltages exist. Or you can use an ohmmeter to check front-to-back ratios. However, with IC field-effect transistors you must be extra careful. Their performance depends on the relative perfection of the insulating layer, if any (some IC FETs do not have an insulating layer), between the gate (equivalent to the base of a bipolar transistor) and the active channel. Should this layer become punctured by inadvertent application of excess voltage to the external gate connection, such as the battery voltage across the leads of an ordinary 1,000-ohm/volt VOM, the damage done is irreversible.

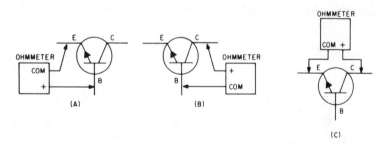

Fig. 4-14. Resistance testing an npn transistor. (A) Testing emitter—base forward resistance; (B) testing collector—base forward resistance; (C) testing collector—emitter forward resistance.

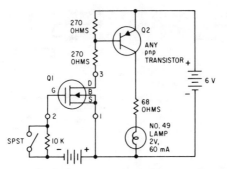

Fig. 4-15. Simple go/no-go test circuit for MOS transistors. *(Note:* Q1 is an n-channel depletion type shown in this example. A p-channel would have the arrowhead pointing outward.) *(Courtesy* RCA Corp.)

The circuit shown in Fig. 4-15 will enable you to determine whether your IC MOSFET is open or shorted. It is a simple go/no-go circuit that will test out-of-circuit n-channel depletion types or p-channel enhancement types. The substrate (**B**) and source (equivalent to the emitter of a bipolar) are connected to terminal 1, the gate to terminal 2, and the drain (equivalent to the collector of a bipolar) to terminal 3. If the MOSFET is a dual-gate type, test each gate separately.

For n-channel depletion types, if the lamp lights with the switch open, then goes off when the switch is closed, the transistor is "good," that is, neither shorted nor open. If the lamp lights whether the switch is open or closed, the transistor is shorted. If the lamp does not light at all, regardless of switch position, the transistor is open. For p-channel enhancement types, the lamp should light with the switch closed and go out when the switch is opened. Otherwise, the test is the same.

Digital (Pulse) Circuits

As with linear circuits, digital circuits also consist of resistors, capacitors, diodes, and transistors. However, here the similarity ends. Digital logic diagrams do not show discrete transistors; they show symbols. These symbols represent functions (AND, OR, etc.), with the functions being performed by combinations of resistors, diodes, and transistors, as shown in Figs. 4-16 and 4-17.

Checking complete digital circuits is far beyond the scope of this book. To check a digital circuit thoroughly you must be very fa-

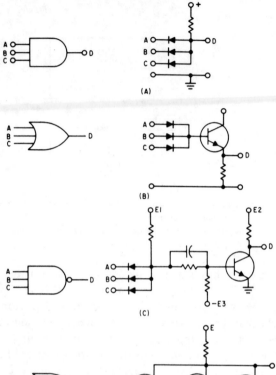

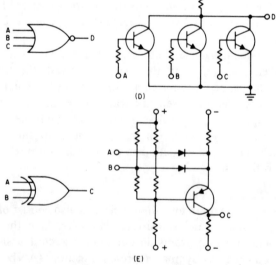

Fig. 4-16. Basic logic gates in circuit form (right) and with logic symbol (left). (A) Three-input AND gate; (B) three-input OR gate; (C) three-input NAND gate; (D) three-input NOR gate; (E) two-input Exclusive-OR gate.

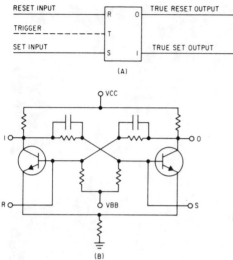

RESET INPUT

TRIGGER

SET INPUT

R 0

T

S I

TRUE RESET OUTPUT

TRUE SET OUTPUT

(A)

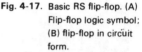

(B)

Fig. 4-17. Basic RS flip-flop. (A)
Flip-flop logic symbol;
(B) flip-flop in circuit
form.

miliar with all the complex logic symbols and the way they inter-
relate in a logic diagram, and you must have the proper test equipment. Testing, therefore, is limited to checking a few of the basic circuits (Figs. 4-16 and 4-17) using the Logic Probe and Logic Pulser
shown in Fig. 4-18.

The Logic Probe lights up near its tip when the tip is touched to
a "high" level. The probe also stretches pulses that are 25 nanosec-
onds or wider, to give a light indication of 0.1 second. The light then
flashes on or off, depending on pulse polarity. For static tests, the
pulser feeds in a pulse, while the probe indicates whether the appropriate high level or low level exists at the proper IC pins.

Testing an AND Gate. Assume the AND gate shown in Fig.
4-16A connected in a DTL (diode-transistor logic) or TTL (transis-
tor-transistor logic) system.

1. Connect the probe to a source of +5 V dc; the band should
   light up.
2. Place the probe tip on any gate input pin. If a "high" level is
   expected at the pin, the band should remain lit; if a "low" level
   is expected, the light should go out. If a series of pulses is expected, such as clock pulses, the band should be lit dimly if the
   expected frequency is 1 MHz or less, and lit dimly or go out
   momentarily if the frequency is above 1 MHz (up to 20 MHz).

Fig. 4-18. H-P Model 10525T Logic Probe and Model 10526T Logic Pulser. *(Courtesy* Hewlett-Packard)

3. If the expected levels or pulse trains appear at the proper input pins, shift the probe to the gate's output pin. Depending upon the logic system used, the band should remain lit for a high ("1"—positive logic, "0"—negative logic), and extinguish for a low (vice versa).

   Testing an OR Gate. Assume the OR gate shown in Fig. 4-16B.
1. Connect the probe to a source of +5 V dc.
2. Place the probe tip at any gate input pin. If the band remains lit, shift the probe tip to the gate output pin. Depending on the logic system, the band should remain lit for positive logic and be extinguished for negative logic. Repeat the procedure for each input pin, each time checking the output in the same manner.
3. The gate can also be checked statically, using both the probe and the pulser. With the pulser also connected to +5 V dc, and

the equipment under test powered down, simply place the pulser tip at any gate input pin and the probe tip at the gate output, as shown in Fig. 4-18. A good OR gate will cause the probe band to go on and off in coincidence with the pulser input. Repeat the process for each gate input, with the probe fixed on the gate output.

Testing a NAND Gate. Assume the NAND gate shown in Fig. 4-16C. The test procedure is the same as for the AND gate, but the result should be the opposite. That is, where a "1" is expected at the AND gate output, the NAND gate should produce a "0."

Testing a NOR Gate. Assume the NOR gate shown in Fig. 4-16D. The test procedure is the same as for the OR gate, but the results should be the opposite. That is, where the OR gate should cause the probe band at the output to go on then off, the NOR gate should cause the band to go off then on.

Testing an Exclusive-OR Gate. Assume the Exclusive-OR gate shown in Fig. 4-16E. The procedure and results are the same as in the third step of the OR gate test.

Testing a Flip-Flop. Assume the simple RS (reset-set) flip-flop shown in Fig. 4-17.

1. Connect the logic probe and logic pulser to a source of +5 V dc.
2. Place the probe tip at the *set output* of the flip-flop, then place the pulser tip at the *set input.* The probe band should remain lit.
3. With the probe still at the *set output,* move the pulser tip to the *reset input.* The probe band should now be extinguished. You can also check the *reset* function the same way, with the probe tip placed at the *reset output.*

Index